Soviet Agriculture

edited by
Harry G. Shaffer

Soviet Agriculture

An Assessment of Its Contributions to Economic Development

PRAEGER SPECIAL STUDIES IN INTERNATIONAL ECONOMICS AND DEVELOPMENT

Praeger Publishers New York London

Library of Congress Cataloging in Publication Data

Main entry under title:

Soviet agriculture.

(Praeger special studies in international economics and development)
Includes bibliographical references.
1. Agriculture—Russia. 2. Agriculture—Economic aspects—Russia. 3. Russia—Economic conditions.
I. Shaffer, Harry G.
S469.R9S644 338.1'0947 77-7512
ISBN 0-03-021976-0

PRAEGER SPECIAL STUDIES
200 Park Avenue, New York, N.Y., 10017, U.S.A.

Published in the United States of America in 1977
by Praeger Publishers,
A Division of Holt, Rinehart and Winston, CBS, Inc.

789 038 987654321

To my children--Bernie, Ronnie, Lennie, and Tanya--in the hope that they will approach all problems with open minds, always ready to expose themselves to a wide variety of views before forming their own opinions on any controversial subject.

PREFACE

More has probably been written on Soviet agriculture, both in the USSR and in the West, than on any other aspect of the Soviet economy. To a great extent, this interest can be accounted for by the importance of the Soviet Union's agricultural sector. It is not merely that food and raw materials there, as elsewhere, are essential for overall economic development; it is also that agricultural output has a considerably greater impact on overall economic performance in the USSR than in the West. Although its relative share has been diminishing, the agricultural sector still accounts for 20 percent of the Soviet GNP, as compared with less than 5 percent in the United States.[1] Moreover, Soviet agriculture holds special fascination for Western observers interested in comparative economic systems because its institutions are so different from those of Western agriculture. Industrial goods, after all, are produced quite similarly in economically advanced socialist and nonsocialist countries (steel, for instance, is produced in the United States and in the Soviet Union by tens of thousands of laborers working together, under the direction and supervision of hired managers); but the structure of Soviet agriculture with its collective farms, its state farms, and its private subsidiary plots is very different from the system of family farms that predominates in the West (and, by the way, also in two of the socialist countries, Poland and Yugoslavia).

The bulk of Western nonsocialist literature is generally critical of Soviet agricultural performance, pointing to relative inefficiency in terms of output per hour of labor time or per acre of land as compared with Western countries, to the slow pace of growth as compared with other sectors of the Soviet economy, and to output all too frequently below the goals set by Soviet planners. But there is also another side to the picture: Soviet agriculture has made progress. Even in a disastrously bad crop year such as 1975, grain output was still well above the average for the early 1960s and in good crop years such as 1973 or 1976, it was more than 50 percent higher; per capita meat consumption, although still only half that of the United States, has doubled over the past 25 years; the consumption of eggs, sugar, and fruit has more than tripled since 1950[2]; and the agricultural labor force, although still several times as high as in the United States, has been decreasing rapidly both in absolute terms and as percentage of the total labor force, from over 75 percent in the pre-Soviet era to barely over 25 percent in 1976.

With its accomplishments and its shortcomings, has agriculture as a whole contributed significantly to the Soviet Union's impressive industrial growth or has it been a drag on the Soviet economy? Are the undeniable shortcomings that Soviet agriculture evidences attributable mainly to unfavorable soil and climatic conditions; could they be accounted for primarily by as yet inadequate capital endowment and training of workers; or should the blame be placed largely on the Soviet system and the socialist and semisocialist structure of its agricultural sector? And does the cure for whatever ills beset Soviet agriculture lie in changing the environment and increasing capital endowment--in other words, in draining swamps, irrigating deserts, expanding production and application of fertilizers and of pesticides, and in putting at the disposal of Soviet farmers more adequate tools and equipment? Could performance be enhanced substantially by better training of farmers or by altering the structure of financial and nonmonetary incentives within the framework of an overall socialist economy? Or is there little hope, lest the system of state and collective farms be abandoned altogether?

On all-important questions such as these, there is no full agreement. Not only do East and West disagree; even Western nonsocialist scholars do not see eye to eye with each other on these issues. Using identical or similar data, they differ in their emphasis, in their interpretations, and in their predictions regarding the future. Probably the majority of Western nonsocialist analysts have a dim view of prospects for significant improvements of agricultural performance under current Soviet economic and political policies; but more optimistic views can be encountered ever more frequently. One recent, semiofficial, U.S. source, for example, ventured the opinion that Soviet efforts could, "in the long run . . . conceivably return the Soviet Union to a major exporter of agricultural commodities."[3]

It is the purpose of this volume to expose the reader to the whole spectrum of views on Soviet agriculture. The book is divided into four chapters, representing divergent Western views--both Marxist and non-Marxist--as well as the official Soviet position.

In Chapter 1, M. Gardner Clark presents a critical analysis of Soviet agricultural performance. Apart from unfavorable climatic conditions and as yet inadequate capital endowment, he sees the problems of Soviet agriculture resulting from a structure and institutions that are inherently inefficient; he contends that this inefficiency of Soviet institutions blunts the effectiveness of resources devoted to agriculture. Professor Clark takes the position that the reasons for the inefficiency of socialized agriculture are, first and foremost, the difficulty of properly supervising large agricultural units, and, second, the difficulty of properly motivating collective and state farmers.

In Chapter 2, Harry G. Shaffer challenges widely held Western beliefs regarding the alleged inadequacy of Soviet agriculture. While fully aware of shortcomings and of the need for further improvements, he yet argues that Soviet agriculture played a major and reasonably adequate role in providing the basis for the Soviet Union's remarkable overall economic growth, and he gives credit to Soviet agriculture for continuous improvements in the diet of Soviet citizens. When a shorter version of this chapter was presented at the 1975 meetings of the Eastern Economics Association, a discussant, David M. Schoonover from the Economic Research Service of the U.S. Department of Agriculture, commented that "the argument . . . by Professor Shaffer has long been overdue," and that Shaffer "makes an important contribution by stimulating us to rethink our accepted beliefs about Soviet agricultural performance."

In Chapter 3, Victor Perlo presents a Marxist analysis of Soviet agricultural development, performance, and prospects. Neither uncritical nor unaware of persisting difficulties, he does not deny that Soviet agriculture has lagged behind the goals of government and Party leaders. Yet, in spite of admitted shortcomings, he finds that Soviet agriculture and the agriculture of other socialist countries that have followed it, have made the most rapid and dramatic progress ever made in agriculture under any social system.

Compiled from a variety of official Soviet sources, Chapter 4 is primarily intended to give the reader a feeling for the official Soviet attitude and position in regard to the structure, the historic development, the achievements, the lingering shortcomings, and the prospects for the future of Soviet agriculture.

Although the Soviets do not deny the persistence of some problems and the need for further improvements, they nevertheless proudly proclaim that "Soviet agriculture has come a long way, overcoming great difficulties and achieving spectacular results in the past 50 years."

NOTES

1. U.S. Central Intelligence Agency, USSR Agriculture Atlas (Washington, D.C.: CIA, December 19, 1974), p. 4; U.S. Central Intelligence Agency, Handbook of Economic Statistics 1975 (Washington, D.C.: CIA, August 1975), pp. 42-43; U.S. Department of Commerce, Bureau of the Census, Statistical Abstract of the United States, 1975 (Washington, D.C.: Government Printing Office, 1975), p. 382.

2. Gertrude E. Schroeder and Barbara S. Severin, "Soviet Consumption and Income Policies in Perspective," Joint Economic Committee, U.S. Congress, Soviet Economy in a New Perspective (Washington, D.C.: Government Printing Office, October 14, 1976), p. 623.

3. Gary L. Benjamin, "Soviet Agriculture," Business Conditions, June 1976, p. 10 (an economic review by the Federal Reserve Bank of Chicago).

ACKNOWLEDGMENTS

I would like to express my gratitude to M. Gardner Clark and Victor Perlo for their valuable contributions to this book, and to the Novosti Press Agency in Moscow and the Soviet Embassy in Washington for giving permission to reprint material from Soviet sources. And I also want to thank Deborah Spector for her valuable editorial assistance; Gayle Robb, Sara Henderson, and Anita Williams for their splendid secretarial work; and Mark Johnson for other office chores connected with the preparation of this manuscript.

CONTENTS

LIST OF TABLES

CONVERSION EQUIVALENTS

Pounds per bushel	
Wheat and potatoes	60
Rye and corn	56
Barley	48
Oats	32

One kilogram	= 2.2046 pounds
One centner or metric quintal	= 220.46 pounds
One metric ton	= 10 centners or 2204.6 pounds
One hectare	= 2.471 acres
One acre	= 0.4 hectare
One kilometer	= 0.6 mile

Metric tons to bushels	
Wheat and potatoes	36.743
Rye and corn	39.368
Barley	45.929
Oats	68.894

Bushels to metric tons	
Wheat and potatoes	.02722
Rye and corn	.02540
Barley	.02177
Oats	.01452

To convert centners per hectare to bushels per acre, multiply by:	
Wheat and potatoes	1.487
Rye and corn	1.593
Barley	1.8587
Oats	2.788

To convert bushels per acre to centners (metric quintals) per hectare, multiply by:	
Wheat and potatoes	0.6725
Rye and corn	0.6277
Barley	0.5380
Oats	0.3587

One metric ton of seed cotton	= 1.562 bales of 480 pounds
One metric ton of ginned cotton	= 4.593 bales of 480 pounds

Soviet Agriculture

CHAPTER

1

SOVIET AGRICULTURAL POLICY

M. Gardner Clark

UNDER STALIN, 1928-53

The Essential Contribution of Agriculture

There is almost universal agreement that the socialization of Soviet agriculture served three very important purposes for the Stalinist regime. The first two were political: the elimination of the political opposition by the conservative peasant class of kulaks, and the substitution of ideologically undesirable private farming with socialized agriculture in the form of collective and state farms. The third purpose was economic: the release of farm labor to the industrial labor force. Most, but not all, outside observers agree that agriculture also performed a fourth important (economic) function by providing capital for industrialization in the form of a net flow of farm products to urban and export markets.

James Millar, following the late Jerzy Karcz, maintains that, at least during the First Five-Year Plan from 1928 through 1932, agriculture was a net recipient of real resources. Millar further believes that from 1928 to 1932 the terms of trade favored agriculture and prices changed significantly in favor of agriculture.* Alec Nove questions Millar's conclusions and believes that the maximum exploitation of the peasantry--in the sense of pumping resources out of

---

*Millar also denies that the labor which moved out of agriculture in the First Five-Year Plan was a contribution to industrialization. I disagree with Professor Millar on this point, especially when one considers the entire period of Stalinist industrialization from 1928 to 1953.

agriculture while providing minimal returns to people and minimal technical inputs such as farm machinery--took place at the end of Stalin's regime, from 1948 to 1953.[1] In any case, Millar has put in doubt my long-standing impression that agriculture did provide a net surplus to the industrial sector.

Observers agree that agriculture failed to meet the production and productivity goals that Stalin had set in his five-year plans. This failure was partly due to agriculture's low priority compared to industry in Stalin's hierarchy of values.

Why Socialized Agriculture Failed to Meet Goals for Production and Productivity

There is much less agreement about other reasons for the shortcomings of Soviet agriculture in the sphere of production and productive efficiency during and since the Stalin era. This chapter is concerned primarily with these shortcomings and what can or might be done to mitigate them.

Lack of Investment in Agriculture

Outside observers agree that Stalin forced the collective farms to sell products at low prices to provide food and raw materials to the nonfarm sectors and for export, to pay for the import of foreign equipment and technology. Furthermore, Stalin was penurious compared to his successors in allocating central investment funds to agriculture.* Stalin's neglect of investments in agriculture--fertilizer, land improvement, irrigation, and so on--is one important reason for agriculture's unsatisfactory performance in production and yields.

Poor Natural Resource Base

A second reason for the poor production record of Soviet agriculture is its relatively poor natural resource base. Erratic Soviet weather has caused dramatic crop failures and enormous Soviet

---

*The central investment that agriculture received from the Stalinist regime was primarily in the form of farm machinery intended to release labor for nonfarm work. Farm mechanization is designed primarily to increase output per worker rather than yield per acre. Mechanization, per se, does little to increase yields, and it increases total output primarily by facilitating the expansion of cultivated acreage.

purchases on world markets. Less well-known are the problems Soviet weather causes in normal or even the best years. Although the Soviet Union has almost three times as much land area as the United States (excluding Alaska) and by far the world's largest and best soil resources, where the soil is excellent the rainfall is inadequate and where the rainfall is (barely) adequate the soil is usually poor.

The Arctic Ocean is one source of moisture, but it is frozen most of the year. Even when the winds blow out of the Arctic, very little moisture precipitates; this is because as the air moves south it usually becomes warmer, increasing its capacity to retain moisture. The Pacific Ocean is another potential source that does provide monsoon-type rains in the Far East, but its effect is mostly limited to the eastern slopes of the enormous mountain wall that cuts off the moist Pacific air from almost all of the USSR. Tibet and the mountains of central Asia cut the Soviet Union off from the moisture of the Indian Ocean. Interior lakes and seas are also negligible sources of moisture. The eastern rim of the Black Sea, where heavy rain falls on the contiguous slopes of the Caucasus Mountains, is an exception. A modest amount of moisture reaches the fertile plains just north of the Caucasus Mountains, making them the best nonirrigated farming region in the USSR.[2]

This leaves the Atlantic, far from the heart of the country, as the source of about 85 percent of total precipitation. Prevailing westerly winds bring moisture from the Atlantic through a narrow gap between the Alps and the Scandinavian mountains, first to Western Europe, then to Poland and the northwestern USSR. The Atlantic winds proceed across the Urals into Siberia depositing a continuously declining amount of moisture all the way to Lake Baikal and Central Asia. American observers find it hard to believe that only about 1 percent of Soviet farmland, mainly the eastern littoral of the Black Sea, lies in areas receiving 28 inches or more precipitation per year. Much of the Western United States and practically all of the area east of Nebraska receive more than 28 inches.[3] Consequently, vast areas of the Soviet Union are too dry for normal agriculture. To make matters worse, variability in rainfall increases with lower annual totals. An average of 12 inches may mean 8 inches one year, 5 inches the next, and 23 inches the following year.

The rainfall pattern in turn has determined the basic character of the soil. Prairies or steppes predominate where rainfall averages 8 to 16 inches. The decomposition of steppe grass has created chernozem or black earth, the most productive soil in the world, rich in organic matter, high in lime and soluble mineral nutrients. The black earth regions of the Ukraine, the southern Urals, and western Siberia are in the zone of only 8 to 16 inches of precipitation. Black earth underlies about half of the cultivated land.[4]

Where the rainfall is heavier, coniferous forests have replaced the prairies or steppe as the original land cover. Decomposing pine needles created the organic matter in podzols, which cover nearly half of the Soviet Union and underlie about one-third of the cultivated land. These clay soils are acid, contain fewer nutrients, and call for extensive application of fertilizers. The podzol zones are found north of the black-earth zones in areas of 16 to 28 inches of precipitation. They include Moscow, Leningrad, the Urals, and most of Siberia. They are bounded on the north by arctic tundra. South of the black-earth zone are brown and desert soils, which are superior to podzol, but inferior to black earth.[5] Soviet agriculture is handicapped by a colossal mismatch between soil and moisture.

In contrast, an area in the United States that has been supplying the Soviet Union with millions of tons of corn and other grains enjoys the best of both worlds. Illinois, Iowa, parts of Indiana, southern Minnesota, Kansas, and so forth, are blessed with magnificent black soil almost as good and deep as Russian chernozem, plus normally abundant rainfall, averaging from 30 to 45 inches. (American black earth was also formed by the decomposition of prairie or steppe grasses. By a fortunate combination of circumstances, forests did not replace the U.S. prairies even though they receive more than adequate rainfall.)

A severe drought in 1946 prompted the Soviet government to announce a comprehensive program for combating drought, to be carried out from 1949 to 1965. The plan, called "Stalin's Great Transformation of Nature," was to retain moisture in the soil, slow the hot dry winds blowing out of the desert, check soil erosion, and raise crop yields substantially. Most of the work was to be done by collective farmers who were to receive shares in the output of crops from their collective farms. Thus, the enormous investment costs were to be borne primarily by the peasants themselves.[6] A symposium of American experts predicted that Stalin's program would fail primarily because the shelterbelts would not normally grow where rainfall was so low. The program was abandoned after Stalin's death in 1953.[7]

The Caspian Sea has been drying for decades. This reduces the fish catch and speeds the creation of deserts to the north and east. Stalin and his successors have made ingenious and dramatic efforts to halt this process and to increase farm production in these areas by irrigation. Stalin planned a canal to divert the Black Sea into the Caspian, but this project was abandoned because the resulting increase in salt concentration killed the fish in the sea of Azov and poisoned the land on both sides of the canal.[8]

To help refill the Caspian, Stalin also planned to redirect the Amu Darya River from the Aral Sea into the Caspian Sea via a

700-mile-long canal. The project was dropped in 1946, revived again in 1950, and then finally abandoned after Stalin's death.[9]

The Volga is the main supplier of water for the Caspian Sea. Scientists in Stalin's day decided that the best solution would be to replenish the Volga by diverting water southward into it from the Pechora and Vychegda Rivers, which now flow into the Arctic Ocean. This requires a canal across the low watershed between the northern rivers and the Kama, a tributary of the Volga. Work was apparently underway at the beginning of World War II. It was revived in 1956 and has been started again more than once since them.[10]

Work is reportedly underway on a 15-year plan to redirect the waters of four great Siberian rivers, the Tobol, Irtysh, Ob, and Yenisei, from the Arctic toward the desert region around the Aral Sea. U.S. government hydrologists are afraid that this could result in slowing the earth's rotation and possibly increase the degree to which the earth wobbles on its axis. British scientists warn that this tampering might hasten the melting of the vast Arctic ice cap, and thus move the climatic zones of the Northern Hemisphere northward with disastrous results.[11] (Arctic explorer Bernt Balchen, in a study of polar climate, claimed that a general warming trend over the North Pole is melting the polar ice cap, which has decreased from 43 feet in depth in 1893 to a current average thickness of only 6 to 8 feet in winter and summer over an area of 5 million square miles.)

Irrigation has been practiced since ancient times in Central Asia, and was rapidly developed toward the end of the czarist regime. By 1913, irrigation networks covered 4.2 million hectares. Stalin's First and Second Five-Year Plans (1928-37) each added about 1 million hectares and the Third Five-Year Plan, interrupted by the war, probably failed to meet its goal of adding 600,000 hectares of irrigated land. Large irrigated areas were abandoned during World War II. Major reconstruction was undertaken after the war and the Fourth Five-Year Plan (1946-50) apparently fulfilled its goals for the development of 1.4 million hectares of newly irrigated land.[12] Stalin continued to favor the development of irrigation (and shelterbelt planning) in his Fifth Five-Year Plan (1951-55), which called for an increase of 30 to 35 percent in the irrigated area. After Stalin's death in March 1953, the construction of irrigation facilities stagnated as Khrushchev promoted a vast program for bringing virgin and idle lands under cultivation. Consequently, the irrigated area increased only 12 percent between 1951 and 1955.[13]

Stalin's actions indicate that he was well aware of and concerned about the mismatch between soil and rainfall. Furthermore, he initiated several large-scale programs to alleviate the shortage of moisture in rich-soil regions. However, Stalinist Russia was very short of capital, and whenever the very large investments for these projects conflicted with higher priority requirements they were curtailed or abandoned.

Khrushchev subsequently developed an interest in irrigation. During the ten full calendar years of his regime, from 1954 to 1963, the area actually irrigated increased by about 1.8 million hectares at a cost of some 2.874 billion new rubles, or an average development cost of about 1,600 rubles per hectare. Near the end of Khrushchev's reign in 1963 approximately 8.5 million hectares (4 percent of the total sown area) were irrigated.[14]

Another tremendous handicap to Soviet agriculture is temperature. Permanently frozen ground underlies more than 3.5 million square miles of the north and east, and the soil never thaws more than a few feet. This is an area about as large as the entire United States (including Alaska). More than 30 percent of the Soviet Union is too cold for agriculture, and U.S. government analysts maintain that an additional 40 percent is so cold that only hardy, early-maturing crops can be grown. This contrasts with the United States, where cold is a limiting factor in only about 20 percent of our territory.[15] A large part of the Soviet Union is suitable for only the earliest varieties of vegetables and grains. Much of the area with sufficient moisture is unsuitable for farming because the growing season is too short and too cold. Once again, Soviet resources are mismatched.[16]

In the southern zones, where temperatures are warm enough for growing a wide variety of crops, moisture is generally inadequate. Almost no region in the entire Soviet Union has the combination of rich soil, adequate rainfall, and warm temperatures found in the U.S. Midwest. The only Soviet area that begins to approach this optimum combination is the Black Sea littoral of the north Caucasus.

Inherent Inefficiency of Collective and State Farms

A third important reason for socialized agriculture's failure to meet the production goals that Stalin and his successors set is the inherent inefficiency of collective and state farms as production units. The poor performance of socialized agriculture in the Soviet Union and Eastern Europe has been amply documented in Western and Soviet studies. Evidence includes comparisons of factor productivity (labor, land, capital) between the USSR and other countries, comparisons of collective versus private plot farming within the USSR, the poor showing of socialist agriculture in East Europe, its virtual abandonment in Yugoslavia and Poland, reports of visiting experts, and endless criticisms by Soviet authorities and experts themselves.

The inefficiency of Soviet agricultural labor was evident when some firsthand comparisons were made between Soviet and U.S. farms. I examined the records of 371 New York State dairy farms and chose for comparison the farm whose operations most closely approximated those of a Soviet farm. In light of the description of the Soviet agricultural resource base, it was learned that the U.S. farm enjoyed two natural advantages. It had the same amount of rainfall in the 1953-54 growing season that the farm near Moscow enjoyed during the entire 1956 crop year (16 inches). This means, among other things, that the U.S. farm could rely much more on pasture for feed. (On the other hand, the Soviet farm was located on the Moscow River and used its water to irrigate.) Second, the U.S. farm's soil was predominantly high-lime glacial till, which is better than the clay podzol soil of the Moscow farm.

Relevant data on the Soviet and U.S. farms are presented in the first two columns of Table 1.1. The basic similarity in operations is evident. The U.S. farm was just about 10 percent larger in terms of the three most important measures: acreage operated, milk cows, and brood sows cared for. On the other hand, the U.S. farm had 10 percent fewer other cattle and 10 percent less sown acreage. The U.S. farm concentrated its agronomy exclusively on feed crops, whereas the Soviet farm devoted almost half of its cropland to vegetables, orchard, and potatoes (which serve primarily as feed, but also for human consumption). The U.S. farm purchased $5,066 worth of livestock feed, but it sold $7,000 worth of wheat.

The Soviet farm outproduced the U.S. one by 50 percent in the primary product of both farms (milk), but the margin of superiority narrowed in terms of milk sold, because the Soviet farm fed a much greater share to its heifers. The U.S. farm sold about 82,542 pounds of meat during the year, or some 16,000 pounds more than its Soviet counterpart. In addition it sold 1,553 pounds of wool from 200 sheep. The Soviet farm kept no sheep. On the other hand, it sold or distributed to its members substantial quantities of vegetables and some apples that were not produced on the U.S. farm.

The Soviet farm had a slight quantitative advantage in capital equipment in the form of tractors and trucks, as well as ten times as many horses. Furthermore, it enjoyed the services of the Machine Tractor Station (MTS) to perform all of its mechanized operations, such as plowing, harrowing, seeding, harvesting, and so forth. But the U.S. farm relied on modern automatic machinery for milking and bulk-handling of fluid milk that the Soviet farm did not possess. At the end of the year, the U.S. farm had on hand $29,478 worth of equipment of all types (both power and other equipment). We have no comparable inventory of the Soviet farm's equipment, but if we add the farm's share of MTS equipment to the equip-

TABLE 1.1

Comparison of Soviet and U.S. Dairy Farms

Items	Soviet Farm	U.S. Farm	Average of 371 New York Dairy Farms
Acreage			
Operated	1,606	1,740	218
Total sown	678	611	125
Feed crops	371	611	109
Vegetables	134	None	14
Potatoes	173	None	Under 1%
Orchard	37	None	Under 1%
Livestock numbers			
Milk cows	90	100	22.3
Other cattle	80	73	17.3
Brood sows	30	32	0.4
Sheep	None	200	4.2
Laying hens	600	200	102
Broilers	None	100	--*
Production in pounds			
Milk produced	882,000	587,524	--
Milk fed to stock	110,250	35,251	--
Milk sold	771,750	552,273	--
Wool sold	None	1,553	--
Meat sold	66,120	82,542	--
Eggs produced (dozen)	5,400	3,600	--
Eggs sold (dozen)	5,400	1,800	1,428
Equipment			
Tractors	7	6	--
Trucks	4	2	--
Automobiles	1	1	--
Horses	30	3	0.4
Labor force			
Men	51	7	1.9
Women	160	None	None
Total	211	7	1.9
Productivity (pounds)			
Milk produced per cow	9,800	5,875	8,439
Milk sold per cow	8,575	5,523	8,051
Milk produced per worker	4,180	83,932	--
Milk sold per worker	3,658	78,896	93,006
Number of cows per worker	0.43	14	12
Crop acres per worker	3	87	65

*Data not available.

Sources: The data on the average of 371 New York dairy farms were taken from L. C. Cunningham, Commercial Farming, Central Plain Region, 1953-1954 (Ithaca, N.Y.: Cornell University Agricultural Experiment Station, January 1957). The data on the U.S. farm were kindly made available by Professor Cunningham from his basic records.

ment on the farm, it would not be surprising if the total value exceeded $30,000.

The labor force on the U.S. farm consisted of its owner-operator, who was 69 years old, his three sons and three hired hands, making a total of seven workers throughout the year. The Russian farm had 211 full-time workers, including the director. The U.S. farm had 14 cows per worker, whereas the Soviet farm employed over two workers per cow. The Americans cultivated 87 crop acres per worker, while the Russians cultivated three acres per farm worker.

In summary, the two farms were remarkably similar with regard to acreage, equipment, livestock numbers, and production, but strikingly different in one respect--the size of the labor force.[17] The differing sizes of the farm labor force also stand out in comparisons of national averages, as shown in Table 1.2.

TABLE 1.2

Farm Sizes in the United States and the Soviet Union

	United States[a]	Soviet Collectives[b]	Soviet State Farms[b]
Acres per farm	351.6	26,508	99,578
Workers per farm	1.9	506	635

[a]1964.

[b]1968.

Sources: Narodnoe khoziaistvo SSSR v 1968 godu: statisticheskii ezhegodnik (Moscow: "Statistika," 1969), pp. 330, 423, 438; U.S. Department of Agriculture, Agricultural Statistics, 1969 (Washington, D.C.: Government Printing Office, 1969), pp. 428-43.

On the other hand, when one compares Soviet and U.S. factories, the respective sizes of the labor force remain within the same general magnitude. For example, Soviet and U.S. steel mills that are equal in physical size and level of production exhibit differences of 25 to 30 percent in numbers of workers employed. In other words, both Americans and Russians are able to operate factories efficiently with thousands of employees, but the Russians have great difficulty operating farms with hundreds of full-time workers, and Americans never even attempt to do so. Why? The answer to this question has not received the attention it deserves.

The answer is not to be found primarily in the basic nature of the incentive systems. Firms in Soviet and U.S. industry and agriculture employ systems that appeal primarily to the self-interest of the individual worker. Why do materialistic incentives directed primarily toward self-centered drives--which work reasonably well in U.S. industry and agriculture and in Soviet industry--work so poorly in the case of Soviet agriculture? I suspect that the answer is to be found in the peculiar difficulties of supervising farm operations. The factory-oriented incentive system in Soviet agriculture has been unable to deal successfully with the unique and difficult problems of supervising large numbers of agricultural workers in physically huge enterprises. (In rare cases, such as monastic farms and the Israeli kibbutz, socialist farms have elicited unsupervised initiative from their workers by appealing to nonselfish, other-directed motivation. However, the Communist states, including China, have had little apparent success with such appeals to their farmers.)

American analysts have found that farm management encounters a number of serious handicaps that interfere much less with factory management. Management supervision and coordination of farm operations are complicated by the sequential nature of the agricultural enterprise. In agricultural production, unlike industrial production, the process typically spans several months with significant time lags between stages (plowing, planting, and so forth). The stages are separated by waiting periods, because the biological processes take time to complete. In factory production many different persons at different places in the plant continuously perform a large number of different functions.[18] In farming, because the stages are strung out over a long time period, relatively few persons must be employed at a given time.

The industrial production process allows virtually complete specialization of workers performing a single task. Farming, however, requires workers to shift repeatedly from one kind of work to another throughout the production season. This calls for considerably greater supervision per man in farming than in factories, where most workers perform essentially the same tasks month after month.[19]

Furthermore, the quality of performance at the beginning of the production cycle is usually not apparent until the harvest several months later, at which time it is generally very difficult (or impossible) to fix individual responsibility for poor performance. To ensure adequate worker performance, it is usually necessary to provide close supervision of farm workers at all stages of the production process, a task which is complicated by the fact that it is very difficult to evaluate the quality of agricultural work except in a very limited number of operations (namely, harvesting).[20]

Because many operations are carried on simultaneously in factory production, the rate of production and quality of any particular operation can be controlled by the flow of materials between operations. When, however, operations are performed one after the other in a time sequence, as in farming, control by means of mechanical control of the flow of materials is not possible. Supervisory control in farming, therefore, cannot be mechanical.* It must be purposeful, depending on one's idea of the amount and quality of output envisaged in later operations. The work pace cannot be controlled.[21]

Since industrial production is typically carried on within a relatively small physical space with specific areas assigned to particular operations, it is possible for a supervisor to keep a close eye on a relatively large work force. The ability of farm managers to provide the degree and quality of supervision mandated by the sequential nature of agricultural production is, however, seriously limited by the physical size of the farm and the dispersion of tasks to be done over a large area. A supervisor must travel back and forth over widely separated parts of the farm to keep abreast of changing conditions, and his ability to coordinate the work is hampered by his lack of knowledge of what is happening in different places.[22] This problem is less serious in intensive types of farming that require a relatively small area, such as poultry raising or fattening beef cattle on feedlots. Spatial problems can be partially overcome, as J. Patrick Madden points out, by improving the communication network within the farm, allowing managers to cover a wider area. This would require the expansion of roads, the introduction of electronic communications, the use of helicopters, and so forth, all of which would entail a high cost for adequate supervision. Even if the farm were willing and able to incur the necessary costs, however, supervision normally would not be as effective as that of industrial managers supervising workers concentrated within the relatively small area of the factory.

In addition to spatial dispersion, two other aspects of farming greatly increase the difficulty of managerial coordination and supervision: lack of uniformity of natural resources and unpredictable behavior of resources and the environment. For example, a large farm with several different types of soil is more difficult to supervise, because some parts require more irrigation or more tillage than others, and the farmer often finds it easier to do the work himself than to be continuously advising a hired man less familiar with the

---

*The implication is not that industrial supervision can be purely mechanical. The distinctions between industry and agriculture are of degree.

soil characteristics.[23] Such farmers find it more efficient to expand their operations by acquiring larger and higher-capacity machines, allowing them to cover more acreage without hiring another farmhand. Unpredictable and sudden changes in the weather require sudden changes in work routine and more supervision. The seasonal nature of farming also, of course, means that during many periods of the year there is little or no work for a permanent labor force.

The problem of supervision in U.S. agriculture has been resolved by eliminating the source of the problem, namely, the permanent hired labor force. The agricultural sector in the United States has experienced dramatic labor-saving technological change in a basic institutional structure dominated by owner-operated ("family") farms, allowing fairly large units (in terms of output or sales) to be operated by a single proprietor and the members of his family, with few--if any--permanent employees. In 1964, approximately 74 percent of U.S. farm workers were family workers, and U.S. farms averaged about 0.5 hired workers per farm.[24] The 1964 Census of Agriculture, moreover, shows that only about 19 percent of all commercial farms, and approximately 60 percent of the largest commercial farms, in the United States employed hired labor on a regular basis, as shown in Table 1.1.[25]

Most farms in the United States are of the owner-operator type in which the problems of supervision are to a large extent nonexistent. Recently a few dozen big corporations have tried to engage in farming by growing and harvesting crops, but with little success. This recent emergence of the U.S. corporate farm, however, does not invalidate our basic premise that supervision is a fundamental barrier to economies of scale in agriculture. Corporate farms are large in terms of total assets and sales, but they typically employ only small numbers of permanent workers. In addition, those areas of agricultural production in which the corporate farm has been most successful--namely, livestock feeding, egg and broiler production, and so forth--are those areas in which highly capital-intensive techniques can be adopted most readily and/or in which production is not spatially dispersed. Thus, in general, the corporate farm has been most successful in the United States in those types of agricultural production in which the problems of supervision are minimal.

Because of the nature of agricultural operations and because supervision is so difficult, it is important that those who perform farm work exercise unsupervised initiative. The worker must adapt to special conditions, shift from one task to another, follow one complex operation from beginning to end, take the machinery to an out-of-the-way spot to do the work, shift and select the tools as needed. Those requirements are difficult to elicit by external controls, such as pay by the hour plus close supervision, or by industrial-type (piece-rate) incentive systems.

The owner-operated farm is ideally designed to elicit unsupervised initiative, because the owner-operator and his family bear directly all the losses and profit from the results of their own work. If rain threatens to damage the hay if it is left in the field, the owner-operator will turn the lights on his tractor and work at night to bring it in--an initiative almost unheard of in Soviet agriculture. In those instances where nonfamily, full-time workers are employed on a permanent basis, the difficulty of supervision is minimized, because the number of such employees is typically very small, and because employees and farmers generally work together on the same tasks.

U.S. agricultural experience seems to suggest that the most effective means of overcoming the unique problem of agricultural supervision is an institutional structure and production technique in which the problem of supervising large numbers of workers is virtually absent. Having minimized the problem of supervision by almost eliminating the permanent farm labor force to be supervised, the profit and loss incentive system can appeal to the self-centered aspects of human nature and motivate the farmer to work hard and efficiently.

In direct contrast to the American experience is the experience of socialist collective agriculture. The Marxian economic tradition stresses practically unlimited economies of scale in agriculture as well as industry. As a result of the ideological underpinnings stressing economies of scale, an industrialization strategy based on expropriation of agricultural surpluses, and a desire to eliminate the political opposition of the kulaks, the Soviet government established very large agricultural enterprises--primarily collective barns--with large numbers of workers per farm. While these enterprises were successful in extracting the agricultural surplus to support rapid industrialization, their performance in terms of production, efficiency, and worker productivity has been very poor.[26] Ignoring the crucial and unique problems of agricultural supervision, the Soviet planners established uneconomically large units while retaining relatively labor-intensive production methods in the agricultural sector--that is, agricultural enterprises in which the problems of supervision should be (and apparently are) at their worst. Soviet farms are physically huge by U.S. standards, and--more importantly--they employ large numbers of permanent workers.

All available evidence indicates that the performance of collective and state farm workers on their household plots has been superior to their performance in the socialized sector. The higher level of incentive on private plots is reflected in the tendencies of collective farmers to minimize their effort on collective work, and to spend excessive time and effort on their private plots. In part, this discrepancy is the result of the fact that the worker's income per unit of labor input is greater, and the relationship between effort and reward is clearer on the household plots than in collective production. However, the greater efficiency of the household sector can also be explained partially by the fact that there is no problem of supervision in private-plot production. In the private sector, it appears that unsupervised initiative--resulting from the clear relationship between individual effort and individual income--is sufficient to generate high levels of worker performance. Attempts to deal with the problems of managing and supervising labor in the collectivized sector have been unable to produce the same kind of initiative, and--as the beleaguered hero of Fyodor Abramov's The New Life, A Day on a Collective Farm illustrates--the peasants have been remarkably effective in evading the restrictions placed upon them by the managers.[27] (Abramov's novella is one of the best of a number of Soviet literary accounts of collective agriculture. The hero is a collective farm chairman who valiantly tries to meet his quota in the face of the bureaucratic rigidity of his superiors on the one hand and the individualism of the farm members on the other.)

In the post-Stalin era, Soviet economic policy has given increasing attention to the perennial problems of the agricultural sector, and policies have been undertaken to raise the level of agricultural output by increasing the level of investment and to improve the wages of collective farmers and state farm workers.[28] These policies--while desirable and consistent with the goal of raising worker performance--are severely handicapped in their attempt to maximize the efficiency of collectivized agriculture because they do not deal explicitly with the problem of supervision. The recent experiments in the Soviet Union with the "link" (zveno) system and the various proposals for subdivision of, and decentralization within, collective and state farms seem to indicate a growing recognition of the unique problems of supervising large numbers of workers in such physically large agricultural enterprises. Closely related to the difficulty of supervising large numbers of farm laborers is the problem of motivating them to work by systems of incentives.

During the first experiments with socialized agriculture after the Revolution, distribution of the proceeds was egalitarian according to the principle "from each according to his ability, to each according to his needs." As late as 1928 a survey showed that distribution in the collective farms was based primarily on the number of workers in a household. It was recognized that the application of this principle did not even insure participation in the collective work, to say nothing of the amount and quality of the work done. Therefore, the Stalinist regime decided to apply piecework incentives to the basic farm operations. The device for estimating performance was called the trudoden, or workday. At first the trudoden was primarily a measure of time worked, but very soon the receipt of workday credits was made contingent upon the type and amount of work done, or the fulfillment of the daily work norm (such as plowing so many acres, and so forth).[29]

All jobs were classified into seven grades. The lowest grade (guards, cleaning women, messengers, and so on) was paid one-half of a trudoden for one day's work, and the top grade (senior tractor drivers, combine operators, and so forth) received two trudodni per day. From 1934, material incentives were intensified by putting tractor drivers on progressive piecework. They were to receive 0.5 trudoden for the first hectare plowed during a shift, 0.7 for the second; 1.2 trudodni for the third and 1.7 for the fourth. Collective farm chairmen were likewise put on incentive pay, depending on the size of the farm and its money receipts.

These incentives could at best encourage quantity, not high-quality work, and the Russians found that the best way to measure quality of work was by results. Regulations introduced in 1938 called for premiums in kind, in cash, or in labor-day credits to be paid for any excess over planned yield and planned livestock productivity. Individual premiums were favored, but sometimes the squad was the recipient. This necessitated that daily records be kept of each individual's work.[30] The director of a metal-working factory in Moscow told me that the problem of supervision in his plant was minimal, because almost all of his workers were paid by the piece and, therefore, did their job well without much supervision. The various material incentive and piecework systems in Soviet socialized agriculture (both collective and state farms) are attempts to motivate Soviet farm workers to work hard and well with minimal supervision. They are bound to remain inadequate because of the nature of farm work. The basic reason is that the quantity and quality of the final output of the socialized farms depends very little on the efforts of any individual worker, but rather on the joint and collective efforts of hundreds of other unsupervisable individuals. There is little way of even knowing, much less controlling, the efforts of all these individuals. Why should an individual worker do a careful, conscientious job plowing a cornfield in May in the expectation that his good work will appreciably increase the total output of corn in July, thereby increasing the value of his labor-day credits or earning him an appreciable premium for yield above plan? He knows that in any case the total yield and his income will depend more than anything else on the weather, the soil, and other factors beyond the control of himself and all the other workers on his farm.

The various incentives, whether trudoden, premiums, or money wage payments, are all designed to elicit unsupervised initiative (work). At the risk of oversimplification, one can say that the individual's efforts will increase output, which in turn determines in part the value of his reward, by only 1/n, where "n" is the number of workers on the farm.

Most U.S. farmers employ no permanent hired labor. Those who do so typically pay a single hired hand a flat rate by the hour, the week, or the month. They make no attempt to devise complex piecework systems of payment, because they are able to supervise their hired hand, controlling both the quantity and the quality of his work.

In addition to incentives, the Stalinist regime developed various means of compulsion, such as penalties for absenteeism and other violations of labor discipline. Private garden plots and holdings of livestock were reduced in size to diminish labor input into private plots that drew farmers away from collective work. More important were the establishment and continual increase in the minimum number of days per season and per year required in collective work.[31]

In addition to the difficulty of supervision, the incentives were limited in their effectiveness by Stalin's persistence in keeping socialized agriculture's real income so low. This was the consequence of his attempt to use the collectives as a device for extracting capital from the peasants. His principal method was the extremely low prices he paid for compulsory deliveries. This meant that the net incomes of collective farms were low, and all the labor-day credits and premiums had to be paid out of the low net income of each collective.

KHRUSHCHEV SHIFTS THE EMPHASIS TO PRODUCTION, 1954-64

In the difficult years following World War II, agriculture contributed to the reconstruction of the Soviet economy in the same way that it had originally contributed to rapid industrialization from 1929 to 1940.

By the time of Stalin's death in March 1953, many of his earlier goals, in whose attainment agriculture had played a key role, were no longer of highest priority. The kulaks had long since been liquidated, socialist agriculture was well-established, tens of millions of ex-farmers and their children were recruited and trained for nonfarm work, and capital was much more abundant than it had been in 1928. The Soviet Union had recovered from the war and become the world's second greatest industrial and military power. Humanitarian values called for long-delayed improvements in the standard of living of Soviet peasants and workers who had sacrificed so much under Stalin. Neither Malenkov nor Khrushchev was disposed to rely on Stalinist terror for control. Both humanitarian values and greater reliance on voluntaristic incentives required higher priority for the previously neglected goal of increased agricultural production and productivity.

Khrushchev's primary concern was agricultural production. He constantly reiterated the theme that "Communist society cannot be built without an abundance of grain, meat, milk, butter, vegetables, and other agricultural products."[32] He intended for the Soviet Union to overtake the United States in the per capita production of meat and milk by the early 1960s.

In the first five years of his regime, Khrushchev was eminently successful in raising agricultural output. From 1953 to 1958, the Soviet official index of gross agricultural production rose by 50 percent and the net production index (as calculated by the U.S. Department of Agriculture) rose by 42 percent.[33] Total inputs into agriculture, especially land and capital, increased significantly during the 1950s. More specifically, inputs of land and capital increased considerably, but labor inputs declined slightly, much less than the declines under Stalin (Table 1.3). Since agricultural output increased even faster than inputs, total factor productivity, especially labor productivity, increased appreciably.[34]

Capital Investment

Total agricultural investment in both state farms and collectives began to increase noticeably in the last two years of Stalin's regime, and even more obviously in 1954 and 1955 under Khrushchev. Investment in agriculture comes from two sources. Under Stalin the greatest share consisted of investments that the individual collective farms made from their own retained earnings. Stalin provided very little through budget appropriations from the government and long-term loans from the state banks.[35]

Under Khrushchev, state investment in agriculture almost tripled, from 985 million in 1953 to 2,700 million rubles in 1957-58. At the same time, investment by collective farms out of their own retained earnings increased from 1,168 million in 1953 to 2,843 million in 1958 and 3,526 million rubles in 1959. Khrushchev's drastic increase in the prices paid by the state for its purchases of farm products made this reinvestment possible. The official index of all farm prices rose threefold from 1952 to 1958 (see Table 2.8) and the prices of items whose production Khrushchev was particularly eager to increase rose even more sharply (for grain sevenfold and for livestock products twelvefold).[36] The prices paid under Stalin were so low that they were considered a form of taxation. Khrushchev raised prices, hoping to transform them into an incentive and reward for increased production and sales.

TABLE 1.3

Agricultural Inputs, 1950-75: Average Annual Employment
(thousands)

Year	State (A)	Collective Farm (B)	Private (C)	Agriculture (D)	Tractors Delivered[a] (thousands) (E)	Total Sown Area[b] (millions of hectares) (F)
1950	3,437	17,600	11,998	43,035	92	146
1951	3,565	26,933	10,614	41,112	?	?
1952	3,693	26,267	10,059	40,019	?	?
1953	4,026	25,600	10,871	40,497	76	157
1954	5,966	25,200	11,482	42,648	?	166
1955	6,041	24,800	12,233	43,074	123	186
1956	5,954	25,700	12,510	44,164	140	195
1957	6,628	24,300	12,486	43,414	148	194
1958	6,005	24,900	12,600	43,505	158	195
1959	5,567	24,500	11,499	41,567	144	196
1960	6,693	22,300	10,901	39,994	157	203
1961	7,496	20,700	11,258	39,454	185	205
1962	7,817	20,000	11,294	39,111	206	216
1963	7,954	19,400	10,862	38,216	239	219
1964	8,168	19,200	10,906	38,274	223	213
1965	8,704	18,900	12,029	39,633	240	209
1966	8,894	18,600	12,291	39,785	276	207
1967	8,836	18,400	11,909	39,145	287	207
1968	8,899	18,100	11,616	38,615	290	207
1969	9,083	17,500	11,156	37,739	304	209
1970	9,180	17,000	11,373	37,553	309	207
1971	9,499	16,500	11,269	37,268	313	207
1972	9,647	16,200	10,995	36,842	313	211
1973	9,885	16,100	10,981	36,966	323	215
1974	10,102	15,900	10,963	36,965	348	216
1975	10,330	n.a.	n.a.	n.a.	370	?

Note: ? = unknown; n.a. = not available.

Sources: Columns A through D from U.S. Department of Commerce, Bureau of Economic Analysis, Estimates and Projections of the Labor Force and Civilian Employment in the U.S.S.R., 1950 to 1990, Foreign Economic Report No. 10 (Washington, D.C., September 1976), pp. 40-41. The figures in column C for private subsidiary agriculture are estimates based on labor inputs calculated by Soviet analysts, and are approximations. For an explanation see pp. 36, 42-43. For estimates of farm labor inputs by various categories of employment and by sector see pp. 37-43, 53-55, 61-63. Columns E and F from various issues of USSR, Ts. S.U., Narodnoe khoziaistvo SSSR and the Report of 1975 Plan Fulfillment. (The total sown area in 1928 was 113 million hectares.)

Land Use and Crop Patterns

Khrushchev had been thoroughly involved in agriculture at the end of Stalin's reign. In his analysis of the situation, in a memorandum to the Presidium at the beginning of 1954, he addressed himself primarily to the task of increasing grain output, which he saw as the root of the agricultural problem Stalin had bequeathed. Grain production in 1953 had failed to cover the increasing demand of the rapidly rising urban population. Shortfalls of production aggravated the heavy burden of high procurements on collective farms and left little to distribute among the peasants. This in turn adversely affected production. Khrushchev recognized that the way out of this dilemma was considerable expansion of grain production.[37] To achieve this, he first authorized plowing grassland pastures and hayfields and planting them with grain. His second method was the traditional Russian one of plowing up virgin land. Between 1913 and 1953 the sown area in the steppes east of the Volga and the Urals had increased from 63 million to 119 million acres. Khrushchev's virgin lands program increased the sown area in the eastern zone by about 90 percent compared to 1953, most of it during 1954 and 1956.

Much of the best land had already been plowed before Khrushchev's new drive started in 1954. But as our description of Soviet agricultural resources has indicated, the biggest handicap to the virgin lands program was the climate: low temperatures, short growing seasons, and, above all, lack of moisture (from 10 to 16 inches average annual rainfall). Furthermore, the lower the average rainfall is, the more it fluctuates from year to year. As a result the yields of grain from the virgin lands were low and variable. Attempts to mitigate the variation and raise the general level of yields were handicapped by questionable practices. According to Lazar Volin, the major obstacle was the virtual single-crop system of spring wheat and its continuous cropping year after year. In his opinion,

> Notwithstanding lower yields, a larger proportion of grain could be procured in these sparsely settled regions of predominantly huge state farms than in the older regions of dense peasant population, requiring larger quantities of grain for human consumption and animal feeding. The new lands thus accounted for most of the increment in the government procured grain and performed a kind of insurance function in some years when weather conditions in traditional regions were unfavorable and yields low. But it is precarious insurance.[38]

Khrushchev's third method of increasing the quantity and quality of agricultural production was to restructure the crop pattern in the traditional producing regions. He wanted to bolster the lagging feed supply so that he could increase livestock production and thus improve the starchy Soviet diet. Inspired by the corn-hog system so successful in the United States, he embarked on a crash program of corn production, expanding its acreage from 11 million acres in 1954 to 44 million in 1955 and to a peak of over 90 million acres in 1962. About 85 percent was harvested for silage, in contrast to the United States, where roughly 85 percent of corn acreage is left to mature and then harvested for grain.[39]

The output of corn silage increased sharply, from 43 million tons in 1955 to a peak of 206 million in 1960. Khrushchev criticized the low yields per hectare, and toward the end he recognized the poor feeding quality of Soviet corn silage. As he put it, "Corn silage without ears is an inferior feedstuff. . . . You talk about tons of harvested corn but actually you were harvesting water. . . . There seems to be a lot of tonnage but nothing to eat."[40] After Khrushchev's ouster, the acreage of corn for silage was cut back drastically.

Low temperatures and short growing seasons over most of the Soviet Union prevent corn from maturing into grain. Nevertheless, in March 1955, Khrushchev proposed that by 1960 the Soviet Union should produce 60 million tons of grain corn, compared to the U.S. 1955 crop of 33 million tons and the Soviet 1953 crop of only 3 million tons. Of course he failed to meet his goal, but he reached a maximum of 17 million tons in 1961. Thereafter output dropped to a low point of 8 million tons in 1965, then remained slightly above that level, rising in 1973 to 13 million tons as the need for this high-energy feed increased.[41] Yields were disappointing, averaging only about 40 percent of U.S. bushels per acre.[42]

In order to increase corn, Khrushchev drastically reduced acreage under high-yielding winter wheat, reliable rye, and oats, which are more suitable for the climate and poor soils of the northern and central regions than corn. Another apparent mistake was his replacement of perennial grasses like clover and alfalfa with corn in the humid, non-black soil zone.[43]

Khrushchev's attempts to expand grain production, especially corn, were part of his desire to improve the Russian diet by increasing the output of animal products. Adequate rainfall and mild climate throughout western Europe and most of the United States favor livestock feeding in pasture and on hay. Inadequate rainfall, harsh weather, and the shortage of grazing land in many important agricultural regions of the Soviet Union necessitate longer periods of stall-feeding with greater requirements for feed grains. Therefore, in order to increase livestock supply, the USSR has to depend even more than Europe and North America on increased supplies of feed grains. Furthermore, animals must be fed several pounds of grain to produce one pound of meat. Khrushchev's goal of catching up with the U.S. per capita meat production required an enormous increase in feed grains. The failure of his corn program was a devastating blow to the attainment of his livestock goals.

Livestock

In May 1957 Khrushchev called for Soviet per capita meat supply to equal that of the United States by 1960 or 1961. According to Khrushchev, the average Soviet citizen ate about one-third as much meat as his American counterpart. He compared live-weight meat production in the United States (17,200,000 metric tons) to 6,500,000 metric tons in the Soviet Union. In carcass weight, however, U.S. production was more than three times that of the Soviets. Furthermore, the Soviet total includes offal, poultry,* game, horse meat, goats, and rabbits and lard, all of which are listed separately in the United States, where the weight of poultry alone was more than 3 million tons. Comparing the consumption of beef, veal, lamb, mutton, and pork, excluding lard, Soviet per capita consumption was not one-third, but between one-fifth and one-sixth of U.S. consumption. The U.S. figure in this corrected comparison covers only civilian consumption, while the Soviet total includes meat for its well-fed armed forces as well as the government's emergency reserve. The figures on meat must be as carefully explained as those on grain.[44]

Lazar Volin has claimed: "Just as there are two parallel roads for expansion of crop production--increased acreage and yield per acre--so there are two avenues for expansions of livestock production: increased number of animals and greater production per animal. The USSR took the first and the easier of these two roads both in crops and in livestock production."[45] As a result, great numbers of poorly fed animals and birds were held on farms long after their useful lifespan, providing low meat, milk, and egg yields. When the available feed is spread thinly among rapidly increasing numbers of animals, it merely keeps the animals alive for long periods of time. Both labor and feed are wasted, producing little additional meat. In the case of hogs, Khrushchev tripled their number between 1954 and the end of 1961, considerably exceeding the U.S. hog population, but obtaining just over half as much pork production as the United States.

---

*Poultry accounted for 500,000 tons of the 1956 Soviet meat total of 6,500,000 tons.

It would have been much more efficient for Khrushchev to have purchased U.S. corn and fattened a much smaller number of hogs instead of his futile crash program to grow feed corn under the handicap of Soviet climate.

Experts from the U.S. Department of Agriculture who observed Soviet agriculture in July 1963 were struck by the high cost and inefficiency of livestock production. The cattle were stall-fed throughout the year, eating roots, tubers, silage, and green feed rather than grain. Visiting Soviet farms in the same year, I observed hand-milking of cows while milking machinery stood unused in the sheds, as well as hand-cleaning of stalls, and so forth. The Americans also observed that inadequate marketing and transportation facilities prevented efficient regional specialization in livestock and crop production. The shortage of packing and processing plants and refrigerated warehouses effectively limited the interregional shipment of livestock products. Inadequacies in the marketing system forced livestock production to be more closely tied to local markets in the Soviet Union than it is in the United States.[46]

U.S. government studies found that among livestock enterprises, dairying was the furthest advanced in the Soviet Union. Milk production was increasing rapidly. Nonetheless, average milk production per cow was only half that of the United States. This was partly due to the fact that Soviet dairy cattle are largely dual-purpose breeds. Practically all beef comes from dairy cattle. Partly because of the lack of rainfall, grazing of livestock is much less widely practiced than in the United States, and the "green chop" system is often used. A succession of crops are cut green and hauled to the barn for feed. This requires much labor to cut and feed an extremely heavy tonnage of feed every day.[47]

Although Khrushchev had increased livestock procurement prices several times from 1952 on, these inefficiencies meant that the prices were still too low to cover average costs of production on collective and state farms. Therefore, livestock and poultry prices increased by another 35 percent in 1962. In order to make livestock production profitable, Khrushchev had to make the price spreads between grains on the one hand and meat, chickens, eggs, and milk on the other considerably higher than in the United States. For example, the U.S. corn-hog ratio usually varies between 12 and 16 bushels of corn as an equivalent in value to 100 pounds of pork. The U.S. delegation in 1963 found that in the better corn-growing regions of the USSR the ratio was double. From 25 to 34 bushels of corn were equivalent in value to 100 pounds of pork.[48] Naum Jasny found similar indications of the very high cost of Soviet livestock products by comparing the prices of feed grains and animal products in the USSR with the same products in Western European countries as well as the United States.[49]

Machine Tractor Stations

Mechanization ordinarily accounts for most of the capital investment in modern agriculture. Stalin created the capacity to manufacture tractors and other farm implements. The slaughter of horses during collectivization created a severe shortage of draft power, which was alleviated when horses were replaced by tractors. State farms maintained and operated their own machinery, but Machine Tractor Stations (MTSs) were assigned a pool of machinery that they maintained and operated to service, for a fee, a number of collective farms. This permitted fuller utilization of scarce capital equipment.

The MTSs were under the direct control of the state, and their monopoly of farm machinery enabled them to become a powerful instrument of both government and Party control over the relatively independent collective farms. The MTSs also served as a conduit for the government to acquire grain and other farm products. Payments in kind by the collective farms for MTS services surpassed compulsory deliveries as a means of grain procurement as early as 1937 and accounted for 47 percent of total grain procurements as late as 1957.[50]

These functions of the MTS became less important to Khrushchev than they had been to Stalin. Khrushchev had treated collective farmers so much better than Stalin that he must have been much less concerned than his predecessor about the need for political and economic control. Furthermore, the increase in the size and reduction of the number of collectives made Party control of each collective easier than before. Also, as indicated, Khrushchev was less interested than Stalin in maximizing collection, was more concerned with total agricultural production, and perceived that the MTSs were interfering with productive efficiency. The MTSs lacked interest in higher yields, soil conservation, timely seeding, and harvesting. Chronic breakdowns of their equipment, inadequate spare parts, and wasteful use of fuel contributed to their high cost of operation. The existence of "two bosses on the land," the collective farm and the MTS, led to conflicts of interest and great difficulty in determining the responsibility for failure to meet planned goals. This further aggravated the difficulty of supervising the large work force. Therefore, Khrushchev abolished the MTSs in 1958 and arranged for the sale of their equipment to the collective.[51]

The effects of the reform failed to meet Khrushchev's expectations. The average volume of work per tractor increased slightly from 1957 through 1961 and then declined through 1964. Complaints about inefficient utilization, poor maintenance, and repair continued as before. The most serious shortcoming, the lack of timely operation of plowing and harvesting, was apparently due to inadequacy in the numbers and quality of machinery. In the USSR in 1965 the acreage of sown and cultivated area per tractor was five times as much as in the United States. Khrushchev had increased deliveries of farm machinery at a very rapid rate between 1953 and 1957, and then sharply decreased the growth of deliveries in the following seven-year period (see Table 1.4).[52]

Incentives and Labor Supply

Stalin promoted industrialization at the expense of agricultural production. Under his regime, farm mechanization made possible the release of millions of young workers for industrialized labor. Under Khrushchev, mechanization permitted the rapid expansion of crop acreage, which in turn was primarily responsible for the increase in agricultural production, particularly from 1953 through 1958. Under Khrushchev, agricultural exodus was slow and irregular. The total average annual labor force engaged in crop and livestock production declined from 40 million in 1953 to 38 million in 1964 (see Table 1.3). Khrushchev was able to stem the tide of farm exodus that occurred under Stalin by substantially improving the incentives for farm work. Khrushchev's revision of earnings and incentives was designed to retain the farm labor force and to elicit greater agricultural output.

It is relatively simple for the government to manipulate the incentives and earnings of state farm workers. The farms are owned and directly controlled by the central government, which sets the prices it pays for their produce, provides capital by direct grants from the budget, hires and fires managerial staff and workers, who are state employees, and directly determines methods of payment and level of wages. If, as is normally the case, state farm operations lose money, operating deficits are covered by subsidies from the state budget. On the other hand, collective farm workers share the fluctuating net income of the collective farm to which they belong.

By setting high quotas and paying low prices for deliveries to the state procurement agencies, Stalin left the collective farms with little for capital investment or payments to their members. Reduced delivery quotas and sharply increased purchase prices between 1955 and 1957 enabled the collective farms to increase both investments and workers' incomes. Khrushchev claimed that total payments to collective farm workers in money and in kind increased from 4.75 billion rubles in 1952 to 8.38 billion rubles in 1957.[53]

In 1958 Khrushchev reorganized the procurement system. First he unified the different types of procuring operations and established fixed quotas for government procurements generally fixed for a period of five to seven years, allowing quotas to be raised

TABLE 1.4

Grain Production, Numbers of Cattle, and Index of Net Agricultural Output, 1928-75

Year	Grain (millions of tons)	Cattle on January 1 (millions)	Net Agricultural Production (1965 = 100)	Year	Grain (millions of tons)	Cattle on January 1 (millions)	Net Agricultural Production (1965 = 100)
1913	86	--*	--	1952	92	59	56
1928	73	67	--	1953	83	57	59
1929	72	58	--	1954	86	56	61
1930	84	51	--	1955	104	57	70
1931	66	43	--	1956	125	59	78
1932	63	38	--	1957	103	61	81
1933	67	34	--	1958	135	67	88
1934	67	34	--	1959	120	71	86
1935	69	39	--	1960	126	74	87
1936	60	46	--	1961	131	76	94
1937	97	48	--	1962	140	82	92
1938	71	51	--	1963	108	87	80
1939	--	54	--	1964	152	85	97
1940	96	48	--	1965	121	87	100
1941	--	55	--	1966	171	93	109
1942	--	--	--	1967	148	97	109
1943	--	--	--	1968	170	97	115
1944	--	--	--	1969	162	96	110
1945	47	44	--	1970	187	95	126
1946	40	48	--	1971	181	99	126
1947	66	47	--	1972	168	102	118
1948	67	50	--	1973	223	104	136
1949	70	55	--	1974	196	106	134
1950	81	58	57	1975	140	109	122
1951	79	57	53				

*Data not available.

Sources: Grain and cattle data are from Narodnoe khoziaistvo SSSR, 1922-1972 and Narodnoe khoziaistvo SSSR v 1974 godu. See also Joint Economic Committee, U.S. Congress, Comparisons of the United States and Soviet Economies (Washington, D.C.: Government Printing Office, 1959), pp. 230-33. The index of net agricultural output is a U.S. government calculation from Douglas Diamond and Constance Krueger, "Recent Developments in Output and Productivity in Soviet Agriculture," Joint Economic Committee, U.S. Congress, Soviet Prospects for the Seventies (Washington, D.C.: Government Printing Office, 1973), p. 335. See also F. Douglas Whitehouse and Joseph Havelka, "Comparison of Farm Output in the US and USSR, 1950-1970," Joint Economic Committee, U.S. Congress, Soviet Prospects for the Seventies (Washington, D.C.: Government Printing Office, 1973), pp. 361-63; David W. Carey, "Soviet Agriculture: Recent Performance and Future Plans," Joint Economic Committee, U.S. Congress, Soviet Economy in a New Perspective (Washington, D.C.: Government Printing Office, October 14, 1976), pp. 595-99.

annually only if increased harvests warranted. In the process prices rose considerably. Next he differentiated prices geographically to alleviate the handicap of farms located in regions with poor soil and climate. He also instituted the payment of advances at a portion of the value of government purchases to help collective farms meet their expenses before harvest time. At the same time, attempts were made to determine the costs of producing various crops, because prices were to be based on average costs of production. (Under Stalin no attempts had been made to discover production costs.) Studies showed that production costs (not counting interest and rent) varied much more than the highest and lowest zonal prices established in 1958. Therefore, the reforms did little to relieve the handicaps of the weaker collective farms. In spite of their imperfections, Khrushchev's reforms transformed the procurement system from a method of agricultural taxation into a system of incentives to encourage much-needed agricultural production. As Volin concludes, "Instead of serving as a mechanism to pump wealth from the agricultural into the nonagricultural sector procurements are becoming [in the mid-1960s] a source of growing kolkhoz revenue and could be used for economic incentives to the workers."[54]

For several years Khrushchev left Stalin's incentive system intact. In the 1950s the workday unit or trudoden system came under critical fire. It lacked uniformity and had become so complicated that it confused the workers instead of inspiring them to greater efforts. The bonus system linked the trudoden variously to output, marketing, encouraging more work during critical seasons, and favoring certain crops and operations over others. More serious was the uncertainty and irregularity of payment, which was a result of the fact that workday units were residuals paid at the end of the season, usually once a year.

Six months after Stalin died the collectives were encouraged to make more advance payments prior to the end of the year. In 1956 the government procurement agencies were ordered to make advance payments for required deliveries from the farms. The resulting improvement in attitudes toward collective farm work encouraged Khrushchev to begin replacing the trudoden system and payments in kind with a guaranteed monthly money wage. Under the new system, work performance is evaluated not in workday units but directly in money terms, as is customary both in industry and state farms.

Khrushchev's first thoroughgoing attempts to introduce regular monthly money wage payments were handicapped by the lack of working capital. In 1957 and 1958 collective farms began systematically to shift to money wages. Party and government officials often urged the shift indiscriminately, forgetting that collective farms, unlike state farms, did not have the state's unlimited resources at their disposal, but had to pay wages out of their net earnings:

> What happens as a result is evident from the example of the Labor Collective Farm in Zagorsk District, Moscow Province. State farm rates were mechanically applied there in 1959; the carryover fund set up on the farm sufficed for only two payments and then monetary payment was buried. There was a great weakening of labor discipline as a result.[55]

By mid-1962 only one-fifth of the collective farms had reportedly adopted the money wage system. The full implementation of this reform was left to Brezhnev.

Private Plots

Tiny private plots have long supplied about 30 percent of Soviet agricultural output, and larger shares of vegetables and livestock products, from only 3 to 4 percent of the cultivated acreage. The private sector is, however, very labor-intensive, utilizing over one-third of the total input of agricultural labor, and it acquires most of its livestock feed from the socialized sector. Most of the produce goes to meet much of the collective farmer's food and feed requirements, and the rest is sold for cash on the high-priced free markets. The private plots normally provide from one-half to one-quarter of the collective farmer's total income. According to Karl-Eugen Wädekin,

> Soviet leaders have had to utilize the contributions of small private producers to solve the food problem. But they would like to be able to discourage private production so that its significance will decline in accordance with their ideological prediction. If only the output of the kolkhozes and sovkhozes could be increased sufficiently, and, more important, rapidly--then most of the problem would be resolved. But this has proved impossible, and so the private sector has been summoned to the rescue and appropriate concessions have been made. This is precisely what happened during the second half of the 1930s, during the Second World War and the early postwar years, after Stalin's death, and again after Khrushchev was deposed.[56]

Following Stalin's death, Beria, Malenkov, and Khrushchev vied in favoring agriculture's private sector. Malenkov reduced the total tax on private plots by 43 percent. Khrushchev ended Stalin's policy of requisitioning large shares of the peasant's private production through taxation, shifting to a regular money rent for the plots, varying with the locality and type of land. Thus the peasant was not penalized but rewarded for increasing his production and sales. Rents were cut in 1953 and 1954 to help peasants without cattle to buy a cow. All tax arrears were cancelled. These and other measures were particularly designed to encourage private livestock production. As a result, private livestock, which had declined in Stalin's last years, increased sharply from 1953 to 1958. The officially reported gross agricultural output of the private sector increased from 14.7 billion rubles in 1953 to 18.0 billion rubles in 1958. The output of the public sector increased even more rapidly, reducing the share of private output in the total from 45 to 37 percent.[57]

Khrushchev's concessions resulted in an immediate and significant rise in the output of the private sector. This alarmed Khrushchev and he tightened the reins again. Starting in 1956 he denigrated the importance of the private plots, reduced their size, and tried to limit the number of livestock. Between 1958 and 1960 Khrushchev's restrictions developed into what Wädekin calls a great campaign against the private sector. He put pressure on the peasants to sell their private livestock to the collective and state farms, and tried to prohibit livestock holdings on the private garden plots of city dwellers. The reduction of grain payments-in-kind to collective farmers and the conversion to money wage payments reduced the feed supply for their livestock. Because state farm workers have much smaller private plots than collective farmers, the expansion of state farms and the conversion of collectives into state farms reduced the private farm sector relatively and absolutely. Moving farmers out of their private individual homesteads with adjacent private plots into enormous, multistoried apartment houses had the same negative impact on private plots.[58]

The decimation of private livestock reduced the availability of livestock products for city dwellers. The cutback in payments-in-kind forced the peasants to buy foodstuffs from state stores and free markets to feed both themselves and their animals. These and other negative effects caused Khrushchev to mute his campaign against the private sector* during the last three or four years of his regime.[59]

---

*The private sector declined from 8.3 million hectares in 1959 to 6.7 million in 1962. Between 1959 and 1963 the private holdings of cattle declined by 19 percent and of sheep and goats by some 3 percent.

Less than one month after Khrushchev's dismissal, the Russian, Ukrainian, and Estonian Republics lifted decrees that had limited the land size of garden plots and livestock holdings of workers and employees. They called for discontinuing the reduction in size of collective farmers' plots, and encouraged them to acquire livestock and procure feed. Khrushchev had restricted the private sector not only for ideological reasons, but also because he believed that reducing time and resources spent in private plots would help the socialized sector. Apparently this attempt backfired, and Khrushchev's successors decided to use the private sector as a stimulus to production in the socialized sector.

> As long as wages paid by the farms, either in cash or in kind, do not rival those paid by the factory, the private sector will remain an attractive inducement for farmers to remain on the farm. And if they remain but cannot work on the private plot until they have fulfilled their quota of work in the socialized sector, the private sector will have fulfilled a basic function useful to the state.[60]

Furthermore, the Brezhnev regime recognizes that the private sector does not simply compete with the public sector for labor. It can utilize workers that cannot be made to work in the socialized sector because they are either too young, too old, or in poor health.[61]

Brezhnev continued to encourage the payment of money wages to collective farmers; in the process, grain payments-in-kind, their principal source of feed, dropped 44 percent in 1968 compared with 1965. The Gosplan calculated that average earnings of collective farmers from private plots decreased from 41 percent of total income in 1965 to about 32 percent in 1970. Both of these trends resulted from Brezhnev's continuation of Khrushchev's policies (converting collective into state farms and moving the collective farmers out of their private homesteads into urban-type apartments). Nevertheless, several studies agree that the private sector seems assured of a reasonably secure existence for some years to come.[62]

Administrative Superstructure

Khrushchev also attempted to increase agricultural output and improve efficiency by making important changes in the administrative organization that controlled agriculture above the level of the basic production units. He vacillated between administrative forms that controlled minute details from the top and organizational structures that gave relative freedom to the farms. He also varied the responsibilities of the Party and the governmental administrative structure, upsetting both groups.

In March 1953, the six separate Ministries of Agriculture, State Farms, Cotton Growing, Agricultural Procurement, and Forestry merged into one Ministry of Agriculture and Procurement. A few months later the Ministries of State Farms and Agricultural Procurement were reestablished. The Ministry of State Farms was swallowed again in 1957. The Ministry of Procurement became the Ministry of Grain Products in 1956 and emerged as a state committee in 1958. We have already described the abolition of the Machine Tractor Stations in 1958. Their supervisory role over the collective farms was taken over by agricultural inspectorates. In 1961 a State Committee for Agricultural Procurements was created to take over all of the duties of the Ministry of Agriculture except research and education. In 1962 both collective and state farms were placed under a completely new system of Territorial Production Directorates, each of which was placed under Party control. The Communist Party was divided into two branches, one supervising agriculture, the other running industry and construction. In 1963 more shuffling of responsibilities for agriculture took place at the top echelon of the Council of Ministers and the Council of the National Economy. Outside observers generally believe that Khrushchev's administrative reforms did little to improve agricultural production and efficiency. His successors apparently agreed, since most of Khrushchev's changes have gone by the board.[63]

Fertilizer

Although he was much more generous than Stalin in directing resources into agriculture, Khrushchev relied primarily on extensive, rather than intensive, methods to increase output (plowing up virgin land, fallow, and grass, and so forth). After 1958 these methods gave diminishing returns, so he turned his attention toward intensification and improvement of yields.

Following a speech by Khrushchev, a circular published in the fall of 1963 explained, "Obviously, further expansion of areas under cultivation will be of less advantage than an increase in the production of mineral fertilizers. . . . Through increased yields we must raise grain production in the next few years by at least 80 million tons. . . . The struggle for increased yields is primarily a struggle for increased production of mineral fertilizers."[64] Under Stalin, the use of fertilizer had been negligible. Khrushchev increased production from 6 million tons in 1952 to 14 million tons in 1960 and 26 million tons in 1964. His goal for 1970 was 80 million tons.[65]

Another aspect of Khrushchev's drive for the "chemicalization" of agriculture was the use of pesticides to control insects, plant diseases, and weeds. The production of these pesticides quadrupled between 1958 and 1965, reaching 103,000 tons. His goal for 1970 was no less than 450,000 tons. These plans called for enormous investments in the chemical manufacturing industry as well as the application of costly fertilizers and pesticides to the land, not to mention the construction of storehouses for fertilizer and the production of machines for loading, unloading, and spreading it.[66]

It was mentioned how the construction of irrigation facilities lapsed during Khrushchev's effort to bring virgin and idle lands under cultivation. The drought of 1963 and the necessity for major purchases from the United States made Khrushchev eager to increase and stabilize the production of grain with very expensive, capital-intensive irrigation systems. In February 1964 the government called for the addition of 12.5 million acres of irrigated area by 1970, compared to only about 4.5 million acres added in the preceding ten years.[67]

Brezhnev's policies have increased agricultural production by concentrating on methods Khrushchev initiated that required capital-intensive investments. The revision of Stalin's approach to agriculture has now come full circle.

BREZHNEV MODIFIES KHRUSHCHEV'S POLICIES

Capital Investment

Khrushchev reversed Stalin's policy of exploiting agriculture for the benefit of industry. Brezhnev went further, allocating an increasing absolute and relative volume of investment to agriculture. The Soviet Union is now investing about five times as much as the United States in agriculture. The U.S. government estimates that this country is investing $8 to $9 billion a year, and that the USSR is investing the equivalent of $45 to $50 billion a year in the agricultural sector.

Soviet agriculture's share of new fixed investment has grown steadily:

Year	Percent
1961-65	19.6
1966-70	23.2
1971-75	26.2
1976-80, planned	26.9

This includes both productive investment as well as nonproductive purposes such as housing. Productive investment in agriculture alone amounts to about 20 percent, compared to less than 5 percent of total investment in the United States.[68] The absolute total invested in 1971-75 was almost three times that of 1961-65. Most of the increase came from state funds to expand the state farm system, to improve land, and to create specialized livestock facilities. Collective farmers' investments, although somewhat less than half of the total, still more than doubled in absolute terms between the first half of the 1960s and 1970s.

Mechanization

The Soviet Union's steadfast devotion to mechanization is indicated by its production of tractors, which increased as follows (in thousands):[69]

1950	1955	1960	1965	1970	1975*
109	163	239	355	459	550

In 1955, the United States produced twice as many tractors as the USSR, but by 1975 Soviet production was double our own and over 10 percent greater than the combined output of the United States, United Kingdom, and West Germany.[70]

Soviet tractor deliveries have exceeded deliveries to U.S. agriculture for almost 20 years. In 1975, the margin was 370,000 to 140,000. One would expect that Soviet agriculture would, therefore, have as many tractors available as U.S. agriculture. It does not. In 1975, the U.S. inventory of farm tractors was 4,190,000 compared to 2,362,000 in the USSR. Soviet tractors do not last long. The annual retirement rate of Soviet farm tractors as a percent of inventory since 1955 has been two to three times as high as the U.S. rate.[71] If the Soviet retirement rate had been as low as the U.S. rate during the post-Stalin era, Soviet agriculture would now have as many tractors as U.S. agriculture.†

---

*The Tenth Five-Year Plan for 1976-80 calls for the delivery of tractors, trucks, combines, and farm machinery in quantities only slightly above those of 1975.

†Earl M. Rubenking calculated that Soviet agriculture would have had 3.2 million tractors in use by about 1971 if the rate of tractor retirements from the park had been kept at about 4 to 5 percent during the preceding two decades, a rate that has characterized U.S. practice

For decades the Soviet press has featured complaints about lack of spare parts for farm machinery; therefore, I was surprised to discover that the production of spare parts for tractors is huge, relative to the United States. In 1974 Soviet spare part production equaled about two-thirds of the value of new tractor output. In comparison, U.S. production of spare parts amounted to a little over one-fourth of the value of new tractor output. David Carey calculates that Soviet spare parts production in 1974 was equivalent to about 350,000 new tractors, whereas U.S. spare parts production was the equivalent of about 64,000 tractors: "This comparison is especially striking in view of the fact that the U.S. park of tractors in agriculture is about 2 million units larger than the Soviet park, and the average age of the U.S. park is about double that of the USSR."[73]

This amazing contrast highlights another comparative disadvantage of Soviet agriculture and its supporting sectors. Carey finds that the exceptionally large need for spare parts stems from both engineering and managerial deficiencies:

> Engineering deficiencies include: mediocre construction quality, mediocre quality of many parts and uneven durability of a significant share of production. Managerial deficiencies include: overproduction of easily made parts and underproduction of more complex ones; faulty distribution which results in receipts of unwanted parts; poor field maintenance; the tendency of farm and tractor repair organizations to overstock parts to be on the safe side; and finally, over-repair or the tendency to replace unworn parts along with those that are worn. The tendency to replace unworn parts apparently has increased in recent years with the development of a large number of assembly-line type facilities for major overhaul work. Since every tractor that enters one of these shops is totally disassembled, technicians are tempted to replace internal parts that normally are hard to get just to be on the safe side.[74]

---

for many years. At a retirement rate of 4.5 percent, the Soviet farm tractor inventory would have exceeded 4 million by the end of 1975 and equaled the U.S. inventory about one year later, assuming 1976 planned deliveries were met.[72]

Ceaseless complaints in the Soviet press stress inadequate field maintenance. From my observation and reading, this is not due to shortage of personnel, but to difficulties in organizing and supervising maintenance, and to the lack of unsupervised initiative in repairing and maintaining this equipment. I recall taking a Soviet agricultural specialist to visit a farm in New York State operated by one man with the part-time help of his teenaged son. When the Russian visitor saw all the well-kept equipment, he wanted to know how many repairmen were in the work force. The farmer smiled and explained that he made routine maintenance and minor repairs on the spot, and his son helped him after school during the winter to make overhauls and major repairs. He got spare parts from the local farm machinery dealer. The Russian found this hard to believe and kept looking around for more workmen.[75] Similar patterns can be found in the deliveries and inventories of farm trucks and combines. For example, in 1960 the USSR produced 59,000 grain combines compared to the 30,000 of the United States. The Soviet lead widened to 102,000 to 21,000 in 1971 and 88,000 to the 28,000 of the United States in 1974. Nevertheless, the Soviet inventory of combines in 1972 was somewhat smaller than that of the United States.[76]

Another often overlooked cost of Soviet agriculture is industry's provision of capital and operating personnel in the form of hundreds of thousands of trucks and tractors to help gather the harvest.*

Fertilizer

Brezhnev also carried out Khrushchev's commitment to increase production by enormous investment of capital in artificial fertilizer. Output of mineral fertilizer increased dramatically (in million tons):[77]

1950	1955	1960	1965	1970	1975
5	10	14	31	55	90

Fertilizer deliveries to agriculture during the Tenth Five-Year Plan (1976-80) are to grow at an average annual rate of 10 percent. This increase will account for over one-half of the planned rise in grain production. Following Khrushchev's lead, fertilizer is to be applied to those areas with adequate moisture but poorer soil.[78]

---

*Agriculture has received between a quarter and a third of Soviet truck production.

I have photographed fertilizer solidified like rocks in large mounds beside Soviet railroads, and seen fertilizer piled unbagged in the open on collective farms. Complaints in the press point to large losses in transportation and storage, and shortcomings in application that rarely occur in the United States, where the enterprise or individual who owns the fertilizer bears the loss for its neglect or misuse. Keith Bush attributes Soviet neglect to lack of direct interest and incentive. In large work groups the link between the correct and timely application of fertilizers, high yields, and higher earnings is tenuous.[79] Such wasted fertilizer is another example of how institutions reduce the effectiveness of the large and growing resources that Khrushchev and his successors have invested in agriculture.

The Brezhnev regime has applied large amounts of lime to neutralize the acid soils of the podzol zone.* Consequently, this zone, which has more rainfall, has increased its output approximately 80 percent, accounting for a large share of the country's total increase in grain production since the early 1960s.[80]

Land Improvement

Brezhnev's expensive investments in irrigation resulted in the gross addition of irrigated land at an annual average of 360,000 hectares during the five years from 1966 to 1970. The additions increased rapidly during the Ninth Five-Year Plan as follows (thousand hectares):[81]

1971	1972	1973	1974	1975
515	784	960	1,090	1,250

The Tenth Five-Year Plan (1976-80) calls for continued expansion at an annual average of 980,000 hectares, higher than the preceding plan period, but below the high point of 1975.

Nobody questions irrigation of traditional crops like rice and cotton, but outside observers suspect that costs exceed the economic benefits of growing grain, and especially wheat, on irrigated land. The investment cost of irrigated land in 1974 averaged between 5,000 and 6,000 rubles per hectare (between $2,650 and $3,200 an acre) to which must be added annual operating costs in the neighborhood of 75 rubles per hectare ($40 per acre). Nevertheless, Brezhnev had expanded the irrigated area under grain to 6.2 million

---

*Lime is applied in conjunction with fertilizer.

hectares by 1975, which compares with 8.5 million acres of total sown irrigated area under all types of crops in 1963. Future plans call for fast expansion of irrigated grain.[82]

Drainage, less expensive than irrigation, is a second method of reclaiming land. From 1966 to 1970, the gross addition of drained land averaged 782,000 hectares per year, and continued at a high level for the next five years (thousand hectares):[83]

1971	1972	1973	1974	1975	1976-80 Plan*
834	837	905	815	1,017	940

Soviet agriculture has a poor record of keeping its drainage and irrigation systems in operating condition. Drain tiles collapse under the weight of heavy farm machinery. Washing to lower the salinity of irrigated land is partially ineffective due to failure to repair and clean the irrigation and drainage networks, and so forth. Hence the rate of retirement is high. During 1966-70, 6 million discarded hectares offset 5.5 million hectares of gross additions to the irrigated and drained area. Once again, institutional inefficiencies blunt the effectiveness of capital investment.[84]

Prices and Procurements

Despite the manifold price increases Khrushchev carried out, when Brezhnev took over he found that most crops and animal products were being produced at a loss or very low rates of profit. Therefore, in March 1965, he raised the basic procurement prices for crops and livestock products by about 12 to 35 percent, and offered a 50 percent premium for above-plan sales of grain. The next major price increases came into effect in 1970, when the 50 percent premium for above-plan sales of grain was extended to meat. As a result of these and other adjustments, the average basic level of prices for all procurements from collective farms in 1972-73 was reportedly 22 percent above 1965. Keith Bush reports that in mid-1975 the prices insured high profits for crops and more modest profits for animal products.

Khrushchev discouraged the farms by the "ratchet effect," frequently increasing state purchasing targets. Brezhnev improved incentives and made planning easier by setting stable purchase targets for five years ahead at levels attained during the preceding five years, with a 50 percent premium for above-plan sales. Procurement

---

*Annual average.

officials' penchant to interfere in the activities of the farm enterprises has diminished since Khrushchev. Apparently, however, "petty tutelage" still persists. For example, the procurement agencies tend to determine supposedly "voluntary" above-plan procurements for each farm.[85]

Soviet Price-Support Subsidies

Under Stalin, prices paid to farmers were so low and retail food prices so high that the state collected the difference in turnover (sales) taxes. Khrushchev turned this large surplus into a small deficit by sharply increasing farm prices. Brezhnev instituted the world's largest agricultural subsidy by continuing to raise farm prices while leaving retail food prices unchanged. Rarely published Soviet figures indicate that the annual subsidies increased from 4 billion rubles in 1966 to a planned 6.5 billion rubles in 1969, and to 9.3 billion rubles in 1971. Constance Krueger has estimated that the total subsidy for price supports on planned 1974 procurement would be 15 to 16 billion rubles ($20 to $21 billion at the official rate of exchange). This expenditure equals 85 to 90 percent of the overt defense appropriation in recent years, or over one-quarter of national income originating in agriculture. The 16 billion rubles excludes other subsidies such as those granted to cover losses of state farms, and so forth. In recent years, for example, subsidies to cover lower prices charged farm producers for machinery, fertilizer, and processed feeds amounted to an additional 1.5 to 2 billion rubles a year.

Keith Bush points out that the USSR "enjoys the dubious distinction of paying the largest subsidy in the world for a still irregular and uncertain supply of inferior foodstuffs. Meanwhile the Soviet housewife has to pay nearly twice as much--at the official rates of exchange--for a 'family foodbasket' of staple foodstuffs as any of her American, British, French, or German counterparts. The comparison in terms of takehome pay for the average industrial worker is even less favorable."[86] About 90 percent of the subsidies are paid for meat and milk, and could be almost eliminated by doubling the retail prices of those products. However, increases of only 30 to 35 percent in retail meat prices reportedly led to riots in the Soviet Union in 1962 and in Poland in 1970.

Incentives and Labor Supply

In 1966, Brezhnev completed the reform of the incentive system Khrushchev had initiated, by giving collective farmers money wage payments modeled on the state farm system. In order to ensure that funds would be available to pay wages, Brezhnev gave first priority to wages in the distribution of net earnings, ahead of savings and reinvestment, which both Stalin and Khrushchev had favored. The State Bank was told to give loans to the poorer farms and those lacking working capital for wage payments.[87] Unless the government is willing to forgive all collective farm debts, interest and repayment charges will reduce the net income available to such farms for reinvestment in plant and equipment and for infrastructures like schools, clubs, roads, to say nothing of bonus payments to the work force. Such charges surely aggravate the consistently reported differences in income among collective farms. Unless the government is willing to subsidize the less fortunate collectives as it subsidizes state farms, the poorer farms will be unable to provide comparable incomes. Unless the government is willing in effect to convert collectives into state farms, collective farmers' earnings will remain a residual claim in the long run.*

Thanks to price increases, loans, and subsidies, collective farms have been able to increase wages and bonuses to their members. From 1966 to 1970 the collective farmers' average earnings grew by 46 percent while state farm workers' earnings increased by only 36 percent. From 1971 to 1975, they apparently grew almost as much again. These increases were greater than those of nonfarm workers and employees. When income from private plots is taken into account, differentials between rural and urban workers may now be comparable to those prevalent in many industrialized countries.[89] Thus, Brezhnev, like Khrushchev, was able to slow the decline in the farm labor force to about 2 million in a decade, compared to 1 or 2 million a year under Stalin. The exodus was minimal compared to the United States, whose farm labor force declined from 9.9 to 4.5 million between 1950 and 1970.

In the Soviet Union, as in all countries experiencing agricultural exodus, the young, male, skilled, and educated led the exodus. Thus there was a deterioration in the quality of the labor force. The Soviets complain about their shortage of young equipment operators and specialists, but the deterioration in the quality of the Soviet farm labor force was nothing compared to what was happening, for instance, in Italy from 1953 to 1974. In the Soviet Union, the farm labor force dropped from about 41 million to 37 million, while in Italy it dropped

---

*State farms have increased their share of output and employment (see Table 1.3) because they are favored in expanding farm acreage, and because many of the weaker collective farms have been consolidated and converted into state farms.[88]

from 8 million to 3 million, and Italy's young male farm operators almost vanished. However, this decimation of its farm labor force did not prevent Italy from increasing her agricultural output about as rapidly as the Soviet Union, despite Italy's worse natural resource base for agriculture.[90]

In April 1975, no less than 2,718,000 tractor drivers and combine operators plus 1,195,000 truck drivers, or a total of 3,913,000, were working in Soviet agriculture. Together these trained specialists almost equal the entire U.S. agricultural labor force (4,390,000 in 1974).[91] (There are practically no full-time tractor and truck drivers in U.S. agriculture.) The problem, thus, is not lack of trained manpower, but wasteful utilization of available manpower. We see again how the inefficiency of Soviet institutions blunts the effectiveness of resources devoted to agriculture.

Organizational Changes

Brezhnev's most noticeable innovation was to stop Khrushchev's continual reorganization of the administrative hierarchy above farm level. Secondly, Brezhnev abandoned Khrushchev's flamboyant and unpredictable crash programs or campaigns.

At the enterprise level, Brezhnev has continued to convert collective into state farms. Many of the new state farms are specialized enterprises that feed enormous numbers of livestock. They are beyond optimal size by Western standards.[92] One should recall that the main limit to size is not the farm acreage, the number of livestock, or the investment in plant and equipment, but the difficulty of supervising large numbers of workers. However, a modern livestock feeding operation or check hatchery resembles a factory more closely than other types of farm operations. Insofar as the Russians specialize and expand "industrial-type" farm operations, these should be easier to supervise and less subject to diseconomies of scale.

Chicken farming is a case in point. The Soviet Union under Brezhnev has adopted modern American methods of poultry production, concentrating and mechanizing the care of chickens in enormous chicken batteries that contain thousands and even hundreds of thousands of hens. At the same time they imported the U.S. system of large-scale breeding of hybrid chickens. As in the United States, egg production has increased rapidly, and labor required per 1,000 eggs has been cut drastically. Since its creation in 1965, _Ptitseprom_ (Chicken Industry USSR) has organized poultry production; by 1975 it was supplying the state with 70 percent of its eggs and 71 percent of its chicken meat.[93] Furthermore, the food processing aspects of these complexes should offer no more problems of supervision than an industrial enterprise.

The Israeli kibbutz and Mao's Chinese commune have shown how to provide year-round employment for farmers, as well as much-needed consumers' and simple producers' goods, by setting up small-scale subsidiary enterprises. Brezhnev has done the same thing with equal success. One reason for this success must be that small industrial enterprises are easier to supervise than agricultural operations. By 1976, over 830,000 persons were working in collective and state farm subsidiary enterprises, producing furniture, household goods, ceramics, cloth, construction materials, and so on. By 1974, the output of these craft industries amounted to over 12 percent of the gross agricultural product.[94]

To satisfy the demands of consumers in modern societies, food is canned, packaged, frozen, and elaborately treated in other ways before reaching the customer. In market economies, cooperatives and other agribusinesses carry out food processing independently of the producing farms. The Soviet Union, which is just getting into this business on a large scale, is organizing food processing as an adjunct of farm production. Since individual farms are too small to supply all the inputs for a modern processing plant, a number of collective or state farms (mostly collectives) get together and operate the processing plants jointly. Like other factories, processing plants are easy to supervise; but I suspect that the organization of a dozen or so collective farms into an enormous producing and processing combine complicates the tasks of coordinating supplies and supervising labor well beyond the point of optimum efficiency.

Construction and repair of farm buildings, livestock feeding, and food processing have increasingly been organized jointly by several collective farms. Starting from 349 in 1956, the number of interkholhoz organizations and enterprises increased steadily to 3,104 in 1964 and 5,733 in 1974.[95]

In a related development, reminiscent of Khrushchev's agrogorods, collective farmers are eventually to be moved out of their primitive villages into urban-type settlements, complete with all amenities and costing in toto 40 to 45 billion rubles. Their distinguishing feature is to be multistoried apartment buildings.[96]

In 1973 the regime instituted Collective Farm Councils at all levels of the administrative hierarchy above farm level. Organizationally this paralleled a development in industry. The main purpose of the organizations was to reduce the inequality among farm incomes that persists despite the adoption of "uniform" guaranteed wages. This goal was to be accomplished with the aid of capital investment funds, by reallocating machinery from farms where it is plentiful to those where it is scarce, and so on. However, collective farm managers objected strongly both to redistribution and to further bureaucratic interference. As a result, the Collective Farm Councils have not proved effective.[97]

In broad perspective, the Brezhnev regime wants to lessen the differences between rural and urban living and raise the efficiency of agricultural production through the "industrialization" of agriculture.[98]

The Unsupervised Link System

For two or three decades, the zveno (link system) has survived in the agricultural system's interstices. This system and various other proposals for subdividing and decentralizing collective and state farms try to solve the unique problems of supervising large numbers of agricultural workers. Advocates of these reforms argue that by making a small group of workers responsible for production on a given piece of land for several years, they will better understand the relationship between individual performance and earnings, and the need for close supervision will be minimal. In other words, these reforms aim to generate the kind of unsupervised initiative that characterizes the private sector.

The link is a primary work group of about 3 to 12 people on a collective or state farm. Normally a subunit of the brigade, it performs a particular operation in one area one day, and another the next. When a link is responsible for the whole production cycle on a particular piece of land, its character undergoes a basic change, becoming more like a private farm. In the chaotic situation after World War II, links were encouraged in order to get production underway when mechnical equipment was in short supply. By 1950, when tank factories had gone back to making tractors and when machinery once more became available, the emphasis shifted from the manual links to the brigade and collective farm, which could put farm machines to better use and cooperate with the MTSs. In 1950, Pravda categorically condemned the links, their leading defenders recanted, and the movement disappeared for the rest of the decade.

Apparently the stagnation of agricultural output after 1958 led to the quiet reintroduction of links on an experimental basis from 1958-59. Following the disastrous harvest of 1963, Khrushchev spoke favorably of an experiment in which the workers were not paid for the fulfillment of work tasks, but for realized volume of production: "At the same time land must be attached to a brigade or a mechanized link. These carry out on this land all the operations--from plowing to cultivating--from sowing--to harvesting. The higher the harvest, the higher the pay for each quintal of produce received. This method raises the responsibility of each person for production. There is no need to send a supervisor to watch the work of a tractor driver. . . ."[99]

Shortly after expressing sympathy for the unsupervised links, Khrushchev was removed. The subsequent failure of the 1965 harvest apparently inspired a press campaign in favor of the links. A conference that autumn reported spectacular results in raising yields and output per man, and reducing costs through elimination of supervisory staffs, while worrying about the link members' much higher earnings. In 1963 Ivan Khudenko started a successful zveno experiment in Kazakhstan whose results were characteristic. When his state farm's 55,000 hectares were divided among 17 links totaling 67 families, grain yield per hectare tripled, output per worker increased twentyfold, and annual income per full-time worker more than doubled. This and other reported results are hard to believe. They read like the comparison of Soviet and American farms.[100]

Support for the links by the Soviet leadership and the press seems to rise with a bad harvest and die down again after a good one on the principle: "Why bother with ideologically tricky new experiments when we have plenty of bread by the old method?"[101] The main ideological objection to the links is the fear of the de facto revival of private farming. The principal short-run political concern is finding suitable employment for those who would be displaced by widespread adoption of links. On the other hand, looking at the economy as a whole, labor shortages hold back overall Soviet economic growth in the nonfarm sectors, which need labor released from agriculture.[102] Therefore, in the long run, Soviet industry would also benefit from adoption of the links.

Aleksandr Yanov is a Soviet journalist and longtime supporter of unsupervised links who emigrated in December 1974. He maintains that the most effective opponents of the links are the administrators and unskilled workers who see themselves becoming superfluous if the normless links turn out to be too efficient. He cites Ivan Khudenko as a case in point. Khudenko conducted his last successful experiment with links in 1972 on marginal land in Kazakhstan. Labor productivity on Khudenko's farm was 20 times higher than on neighboring farms. In January 1973, Khudenko was arrested on trumped-up charges of trying to steal 1,000 rubles from the state. The next year he died in prison.[103]

Between 1970 and 1972, the supporters of the unsupervised links could not generally publish their views. In 1973, the tabu was lifted, and an advocate of the links commented: "The sooner we abandon the idle idea that we can produce a bumper harvest by having squads of students and soldiers wandering around the countryside singing merry songs, the sooner we will get around to putting all plots of land under the control not of collective or state farm offices but of real live corn growers, who will manage the land accordingly to their wishes and interest, . . . who will know what they are about."[104]

In 1973 the link system suffered a severe blow when G. I. Voronov, Premier of the Russian Republic and the link's most powerful advocate, lost his job and his seat on the Politburo. The disastrous harvest of 1975 reinvigorated the link system. In the drought of 1975, there were reportedly no failures in the fields of the "normless teams." In March 1976, the leading agricultural centralizers, Polyansky and Matskevich, were ousted and V. K. Mesyats was named Minister of Agriculture. Mesyats had been associated with Voronov. Three months later a decree of the Stavropol Region announced that they had allocated all arable land in this rich farming region of the North Caucasus, the best in the USSR, to "normless teams" or unsupervised links.[105]

From many viewpoints the link system and the moves toward "industrial centralization" contradict each other. But the contradictions disappear if one ignores the objections of the ideologues and the bureaucrats, who fear loss of their power and their jobs. It would be perfectly possible to organize the agricultural tasks of, for example, a large food-producing and processing combine on the basis of independent, normless links. Each link could be assigned land and machinery on long-term contracts under which it would provide raw material to the processing combine. Perhaps someday such an experiment will be given a fair trial.

Output and Productivity

During the five years following Khrushchev's ouster, agricultural output increased substantially. Soviet official data indicate that gross agricultural output during the Eighth Five-Year Plan (1966-70) averaged 21 percent more than during the preceding five years. This was achieved, however, at a cost of enormous increases in capital inputs. During 1966-70 total capital investment increased by 69 percent, new irrigation by 180 percent, new drainage by 221 percent, deliveries of tractors by 34 percent, trucks by 72 percent, combines by 22 percent, and mineral fertilizer by 104 percent.[106] Bad weather lowered the growth rate of gross output during the next five years (1971-76) to 13 percent, as compared with the preceding five-year average. But massive material inputs increased at almost the same high rates as before.* At the same time, the agricultural

*Total capital investment during 1971-76 increased by 60 percent, newly irrigated acreage by 221 percent, new drainage by 16 percent, tractor deliveries by 16 percent, trucks by 53 percent, combines by minus 4 percent, and mineral fertilizer by 66 percent.

labor force declined from 38 million in 1964 to 37 million in 1974. Furthermore, Brezhnev stopped the rapid increase in cultivated area that Khrushchev had undertaken, and stabilized it from 1964 to 1974 at close to 213 million hectares (see Table 1.3).

The increased real earnings per man and the increased flow of consumer goods and services to farmers have gone along with the modest increase in output since 1964. Higher agricultural procurement prices made possible higher volume of reinvestment and collective farmers' earnings. Unwillingness to pass these higher costs on to the consumer resulted in enormous farm price support subsidies.

The Tenth Five-Year Plan (1976-80) promises a continuation of the more rapid increase of inputs than outputs. Gross agricultural output is to average 16 percent more than the average annual output of the preceding five years. Deliveries of mineral fertilizer are to increase by 52 percent. Total capital investment and deliveries of machinery are planned at about twice the rate of increase of output, but the excess of inputs over output is to be much smaller than it was during the preceding decade.[107]

Agriculture will maintain its priority among resource claimants during 1976-80, with its share of new fixed investment rising slightly to 26.9 percent, but the yearly growth of investment will decrease sharply from 9.5 percent during 1971-75 to only 3.5 percent. This reflects a general slowdown in planned investment throughout the economy. The expensive land-improvement program is to drop to a level somewhat below that of 1971-75. Brezhnev seems to be counting on above-average weather conditions, and his new plan has gotten off to an auspicious start. In October he announced that more than 216 million tons of grain had already been harvested and hinted that 1976 might even surpass the previous record grain harvest of 223 million tons in 1973.[108] By January 1977, the Soviet Union's 1976 grain crop was indeed reported at a record 223.8 million tons.*

Throughout this chapter, I have relied primarily on Soviet gross output data. In an effort to circumvent their limitations, U.S. government analysts have prepared an index of net agricultural production (Table 1.4). It is based on the physical output of 41 crops and animal products weighted by average prices that all producers received in 1970, adjusted for changes in inventories and for intra-agricultural uses of farm products. The best way to appreciate the limitations of Soviet production data is to examine the U.S. index, its 41 components, and the comprehensive explanation of this measure of agricultural production.

---

*AP report from Moscow, January 5, 1977.

Allowances can be made for Soviet overestimates of grain and meat. However, certain problems of the Soviet data are insoluble--for example, milk production. In the first place, the Russians feed a very high proportion of their milk to calves, but how can one verify reports of how much milk a calf drinks? Second, farm households consume a high proportion of the remainder. The Russians themselves have no comprehensive data on the amount of milk sold off the farm. Only a limited portion of milk produced in the USSR is marketed (the rest remains on the farms to feed rural households or animals), so the bulk of milk production must be estimated from surveys. Moreover, much of the Soviet data on milk marketings is actually a calculated milk equivalent of marketed cream. Thus, any understatement of the milk equivalent conversion factor for cream results in overstatement of total milk marketed and, in turn, total milk output.[109]

Examination of Table 1.4 reveals a creditable, but fluctuating and modest, increase in output under Brezhnev's regime from 1965 through 1975. Soviet economists are concerned about diminishing capital productivity in agriculture. One of them calculated that during 1951-60, for each ruble's worth of added capital stock, annual gross output of agriculture increased by 0.50 of a ruble, but during 1961-72 by only 0.32 of a ruble.[110] Despite this crop in capital productivity, Soviet economists believe that Soviet agriculture will have to continue to receive more than its share of investment until it reaches the position of U.S. agriculture, where fixed capital and energy per worker surpass industry and other sectors by large margins. One Soviet economist believes that "industrialization" of Soviet agriculture will require quadrupling productive capital stock and increasing eightfold the share per worker. He estimates the total investment required to be 800 to 850 billion rubles. Wädekin points out that at the 1976-80 planned growth rate of investment (30 percent every five years), the required total will be reached at the end of the Sixteenth Five-Year Plan in the year 2010. He claims this "would make agriculture a severe handicap for non-agricultural investment and growth capacity in the Soviet Union for almost another half a century."[111]

Comparative Advantage, Specialization, and Trade

The principle of comparative advantage applies to Soviet economic development, even though Soviet leaders ignore it. In its classical form, the theory of comparative cost says that each country should concentrate on the production of goods for which it is relatively better suited on account of natural resources, capacities of labor force and economic institutions, and supply of capital. It should exchange its surplus of such goods with other countries for goods which it is relatively less suited to produce. Appropriate specialization and exchange of goods leads to a larger output and wealth than self-sufficiency or inappropriate specialization.[112] We can determine appropriate specialization and exchange by comparing costs of production, provided the country's prices are based on opportunity costs, which can be translated into world prices through equilibrium exchange rates. Unfortunately, internal Soviet prices and ruble exchange rates make such comparisons extremely difficult, if not impossible. Therefore, we have to estimate potential benefits from trade by evaluating the supply and productivity of the factors of production.

Soviet and Western economists and engineers would probably agree that the Soviet iron and steel industry has natural resources, managerial and technical know-how, trained labor forces, and capital equipment far superior to Soviet agriculture. From a purely economic point of view, it would be to the Soviet Union's advantage to export more iron and steel or other products in which it has a comparative advantage, and to import farm products. The Soviet Union, however, has steadfastly attempted to export farm products, or at least to be agriculturally self-sufficient.

To meet its steadily growing demands for food, the USSR has tried to overcome its lack of rainfall through enormously expensive river diversion and irrigation projects, to improve its poor soil with artificial fertilizer, to increase its low yields by cultivating 250 million additional acres, and to improve its inefficient farming institutions by encouraging and retaining on the land enormous labor forces, as well as managerial and supervisory staffs. Thanks to very large and growing inputs of land and capital, agriculture has met its production goals partially, belatedly, and intermittently, with low and diminishing returns on its investments.

Khrushchev and Brezhnev have tried a variety of organizational and administrative changes and incentive systems during the last 25 years. It is hard to imagine a promising reform that they have not already tried. The experiments with normless links have been by far the most successful, because they come to grips directly with what seems to be the most important reason for the inefficiency of socialized agriculture: the difficulty of supervising a large labor force. If the regime could overcome ideological and political objections and adopt the unsupervised link system as the basic production unit on collective and state farms, it might be able to counteract Soviet agriculture's comparative disadvantage.

However, coping with agriculture's limited natural resources base will continue to be very expensive. Soviet development could be more economical if it would follow the precepts of comparative advantage.

Within the economy, planners have appropriately specialized in supplying protein for the Russian diet. The Soviet Union has a big comparative advantage in fishing compared to animal husbandry. To date it has enjoyed almost free access to the world's best fishing grounds, which it exploits with a fishing fleet that in 1975 totaled 5,937,000 gross tons. This is almost twice as large as the combined fleets of all the leading fishing nations in Western Europe, plus Japan and the United States.[113] And the Soviet fish catch in 1975 was two-thirds the weight of that year's high meat production.*

Fishing, which radically differs from farming, goes on in big floating factories. More than half the Russian tonnage consists of ships over 10,000 tons and accounts for 88 percent of the world's fishing fleet in this category.[115] The reported size of fishing fleets varies, but all sources agree on Russian dominance in both total tonnage and large ships. The advantage of large fishing ships with hundreds of workers is the relative ease of supervision in the compact confines of a ship.

Inability to expand agricultural production fast enough to meet growing demand, combined with recent crop failures, has forced the USSR to import farm products as a short-run expedient. Comparative advantage indicates that they should do this in the long run. Under Brezhnev, imports of meat and meat products have fluctuated as follows (1,000 tons):[116]

1965	1966	1967	1968	1969
252	133	58	60	76

1970	1971	1972	1973	1974
165	225	131	129	515

Meat exports have been negligible, but Russia and the USSR have traditionally been net exporters of grain. At the end of Khrushchev's regime and the beginning of Brezhnev's (1964-66), there was an unusual grain import surplus. A brief export surplus existed from 1967 to 1971, followed by massive net imports in 1972-73. The data on net imports and exports of grain are as follows (in million metric tons):[117]

---

*The fish catch was live weight totaling 10,300,000 tons. Meat production was carcass weight including horsemeat, rabbit, poultry, offal, and lard.[114]

	1964	1965	1966	1967	1968
Net imports	5.2	1.8	4.2	--	--
Net exports	--	--	--	3.8	4.0

	1969	1970	1971	1972	1973
Net imports	--	--	--	10.2	18.3
Net exports	7.1	3.8	5.6	--	--

In the crop year 1974-75, following the excellent harvest of 1973, the Soviet Union cut its imports back to equal exports at 5 million tons. The following year Soviet net imports jumped to 25 million metric tons. From July 1972 to July 1976 the USSR imported 64 million tons of grain, primarily from the United States, and is expected to import an additional 13 million tons by July 1977.* In July 1975, George Meany became annoyed by the USSR's disruptions of the world grain market and its impact on the American consumer, forcing President Ford to negotiate the long-term agreement for regular purchases of 6 to 8 million tons of grain per year.

Comparative advantage indicates that recent imports of meat and grain, particularly feed corn, should become part of the long-term pattern of Soviet development. By importing feed grains and meat, both labor and investment could be released from agriculture to industry, where they are much more productive. The present Soviet pattern of development is expensive, requiring enormous investments in both agriculture and industry. Because additions to the industrial labor force are hard to obtain, industrial output has to be increased by investments to raise output per man. It would be less expensive to transfer to industry labor and other resources now being poured into agriculture. The cost would be the exportation of products like iron and steel, timber, oil, coal, or any others in which the Soviet Union has an advantage compared to corn and meat to pay for food and feed imports. The principal disadvantage would be the loss of complete self-sufficiency in food. This loss could be mitigated by importing only enough grain and meat to cover a portion of the desired future improvements in the Soviet diet. The disadvantage could be further mitigated by increasing Soviet grain reserves.

---

*Recent Soviet imports and exports of grain were as follows (in millions of metric tons):[118]

	72/73	73/74	74/75	75/76	76/77 (estimated)
Exports	1.7	5.9	5.0	0.5	4.0
Imports	21.8	11.5	5.2	25.6	13.0

Such considerations seem so obvious that they may have been taken into account and deliberately rejected. The world's second most powerful nation apparently feels that it can afford the luxury of improving its people's diet while relying on its own limited agricultural resources and institutions.

NOTES

1. James R. Millar and Alec Nove, "A Debate on Collectivization: Was Stalin Really Necessary?," Problems of Communism, July-August 1976, pp. 49-62.

2. George B. Cressey, Soviet Potentials, a Geographic Appraisal (Syracuse, 1962), pp. 39-65; A. A. Borisov, Klimaty SSSR (Moscow, 1948), especially his "Klimaticheskaia karta SSSR," a large folding map.

3. W. A. Douglas Jackson, "The Soviet Approach to the Good Earth: Myth and Reality," in Soviet Agriculture and Peasant Affairs, ed. Roy D. Laird (Lawrence, Kansas: University of Kansas Press, 1963), p. 178.

4. Cressey, op. cit., p. 55; George Cressey, Asia's Lands and Peoples (Syracuse: Syracuse University Press, 1963), p. 624.

5. See also J. S. Gregory and D. W. Shave, The U.S.S.R., A Geographical Survey (New York: John Wiley, 1944), pp. 102-20.

6. Harry Schwartz, Russia's Soviet Economy (New York: Prentice-Hall, 1954), pp. 344-46.

7. See Land Economics, November 1949, pp. 333-64.

8. For recent developments see "The Azov Sea Restoration Project," Radio Liberty Dispatch, March 14, 1975.

9. Schwartz, op. cit., pp. 349-51.

10. Paul Wohl, "Caspian Still Shrinks," Christian Science Monitor, January 17, 1967.

11. Christian Science Monitor, December 17, 1969; March 16, 1970; and June 8, 1972. See also U.S. Central Intelligence Agency, USSR Agriculture Atlas (Washington, D.C.: CIA, December 1973), p. 25.

12. Naum Jasny, The Socialized Agriculture of the USSR (Stanford: Stanford University Press, 1949), pp. 483-84. Also Schwartz, op. cit., pp. 344-51.

13. Central Statistical Administration, Sel'skoe khoziaistvo SSSR (Moscow, 1960), p. 259.

14. Gidrotekhnika i melioratsiia, no. 4 (Moscow, 1964): 23; Vestnik statistiki, no. 5 (1964): 93.

15. U.S. Central Intelligence Agency, op. cit., p. 8. See also Cressey, Soviet Potentials, op. cit., pp. 44-46.

16. U.S. Central Intelligence Agency, op. cit., pp. 8-9.

17. For more details see M. Gardner Clark, "Productivity and Incomes on a Soviet Dairy Farm," ILR Research, Fall 1958, pp. 3-9.

18. J. Patrick Madden, Economies of Size in Farming (Washington, D.C.: Economic Research Service, U.S. Department of Agriculture, 1967), p. 9.

19. Ibid., p. 10. The material on which this discussion is based first appeared in M. Gardner Clark and M. E. Bradley, "Supervision and Efficiency in Socialized Agriculture," Soviet Studies, January 1972, pp. 465-73. I am grateful to the editors of Soviet Studies for permission to use the material.

20. See Michael E. Bradley, "Wage Determination and Incentive Problems in Soviet Agriculture" (Ph.D. diss., Cornell University, 1967), p. 225.

21. John M. Brewster, "The Machine Process in Agriculture and Industry," Journal of Farm Economics, February 1950, p. 74.

22. Madden, op. cit., p. 12.

23. Ibid., pp. 11-12.

24. U.S. Department of Agriculture, Agricultural Statistics 1969 (Washington, D.C.: USDA, 1969), pp. 428, 443.

25. 1964 Census of Agriculture (Washington, D.C.: U.S. Department of Commerce, 1968), vol. 3, pt. 2, p. 14; Dan Cortz, "Corporate Farming, A Tough Row to Hoe," Fortune, August 1972, pp. 134-75. See also John M. Brewster and Gene Wunderlich, "Farm Size, Capital and Tenure Requirements," in Adjustments in Agriculture: A National Basebook, ed. C. F. Christian (Ames, Iowa: Iowa State University Press, 1961), chap. 8.

26. Michael E. Bradley, "Marxism and Soviet Agricultural Problems," in Comparative Economic Systems, ed. Jan S. Prybyla (New York: Appleton-Century-Crofts, 1969), pp. 89-95; Bradley, "Wage Determination," op. cit., pp. 267-68.

27. Fyodor Abramov, The New Life, A Day on a Collective Farm, trans. David Floyd (New York, 1963). See also Alex Nove, "Peasants and Officials," in Soviet and East European Agriculture, ed. Jerzy F. Karcz (Berkeley, Calif., 1967), pp. 57-72.

28. See Bradley, "Wage Determination," op. cit.; Jerzy F. Karcz, "The New Soviet Agricultural Programme," Soviet Studies, October 1965, pp. 162-97.

29. Jasny, op. cit., pp. 401-03.

30. Ibid., pp. 403-10.

31. Ibid., pp. 391-400.

32. Lazar Volin, A Century of Russian Agriculture (Cambridge, Mass.: Harvard University Press, 1970), p. 331.

33. Ibid., p. 342, Table 3.

34. Douglas Diamond and Constance Krueger, "Recent Developments in Output and Productivity in Soviet Agriculture," Joint Economic Committee, U.S. Congress, Soviet Prospects for the Seventies (Washington, D.C.: Government Printing Office, 1973), pp. 317-39; D. Gale Johnson, "Agricultural Production," in Economic Trends in the Soviet Union, ed. Abram Bergson and Simon Kuznets (Cambridge, Mass.: Harvard University Press, 1963), pp. 216-18.

35. Volin, op. cit., pp. 347-50.

36. Ibid., p. 384.

37. Ibid., pp. 482-83.

38. Ibid., p. 493.

39. Ibid., pp. 496-99.

40. Ibid., p. 499.

41. U.S. Central Intelligence Agency, The Soviet Grain Balance, 1960-73 (Washington, D.C.: CIA, September 1975), p. 7.

42. Richard Bell, "How Soviet Agriculture Compares with Ours," U.S. Department of Agriculture, Foreign Agriculture, September 1961, pp. 6-7.

43. Volin, op. cit., pp. 499-508.

44. Paul Wohl, "Meat Shortages Hamper Soviets," The Christian Science Monitor, June 25, 1957, p. 5. See also Johnson, op. cit., pp. 231-33, where he raises questions about Soviet data on meat and milk.

45. Volin, op. cit., p. 509.

46. U.S. Department of Agriculture, Soviet Agriculture Today, Report of the 1963 Agriculture Exchange Delegation, Foreign Agricultural Report No. 13 (Washington, D.C.: USDA, 1964), pp. 20-26.

47. Ibid.

48. Ibid., p. 22.

49. See Naum Jasny, "The Failure of the Soviet Animal Industry," Soviet Studies, October 1963, p. 207.

50. Volin, op. cit., pp. 444-58.

51. Ibid., pp. 458-67.

52. Ibid., pp. 471-73.

53. U.S. Department of Agriculture, Soviet Agriculture Today, op. cit., p. 39.

54. Volin, op. cit., p. 404.

55. Ye Lazutkin, "Increase the Material Stake of Collective Farmers in Communal Production," Partinaia zhizn, no. 7 (April 1961); translated in Current Digest of the Soviet Press, vol. 13, no. 19, p. 16.

56. Karl-Eugen Wädekin, The Private Sector in Soviet Agriculture (Berkeley, Calif.: University of California Press, 1973), p. 3.

57. Ibid., pp. 52, 60.

58. Ibid., pp. 247-304.

59. Ibid., pp. 304-14; John W. DePauw, "The Private Sector in Soviet Agriculture," Slavic Review, March 1969, pp. 63-71.

60. DePauw, op. cit., p. 71.

61. Wädekin, op. cit., p. 331.

62. Keith Bush, "Soviet Agriculture: Ten Years Under New Management," Radio Liberty Research Supplement, May 23, 1975, pp. 28-31. See also Wädekin, op. cit., pp. 340-43; DePauw, op. cit., p. 71; Paul Dibb, Soviet Agriculture Since Khrushchev, An Economic Appraisal, Occasional Paper No. 4 (Department of Political Science, Australian National University, 1969), p. 31; and Stephen Osofsky, Soviet Agricultural Policy (New York: Praeger, 1974), p. 169.

63. For a thorough description and analysis see Howard R. Swearer, "Agricultural Administration under Khrushchev" and a commentary by Jerzy F. Karcz in Soviet Agriculture and Peasant Affairs, ed. Roy D. Laird (Lawrence, Kansas: University of Kansas Press, 1963), pp. 9-50.

64. Pravda, October 6, 1963.

65. Volin, op. cit., pp. 364-68.

66. Pravda, November 21, 1963.

67. Volin, op. cit., p. 372.

68. See David W. Carey, "Soviet Agriculture: Recent Performance and Future Plans," Joint Economic Committee, U.S. Congress, Soviet Economy in a New Perspective (Washington, D.C.: Government Printing Office, October 14, 1976), pp. 585, 590. See also the testimony of Douglas Diamond in Joint Economic Committee, U.S. Congress, Allocation of Resources in the Soviet Union and China (Washington, D.C.: Government Printing Office, 1976), p. 43.

69. For a thorough analysis and comparison with the United States see Earl M. Rubenking, "The Soviet Tractor Industry: Progress and Problems," Joint Economic Committee, U.S. Congress, Soviet Economy in a New Perspective, op. cit., pp. 600-619. See Carey, op. cit., p. 586.

70. U.S. Central Intelligence Agency, Handbook of Economic Statistics 1976 (Washington, D.C.: CIA, September 1976), p. 118.

71. Ibid., p. 618. See also Diamond and Krueger, op. cit., p. 321.

72. Rubenking, op. cit., p. 609.

73. Ibid., p. 606.

74. Ibid., pp. 606-07.

75. See U.S. Central Intelligence Agency, Handbook of Economic Statistics 1976, op. cit., p. 118, and Bush, op. cit., p. 11.

76. See Diamond and Krueger, op. cit., p. 321; Bush, op. cit., p. 13.

77. USSR, Tsentral'noe statisticheskoe upravlenie, Narodnoe khoziaistvo SSSR v 1974 godu (Moscow, 1975), p. 183; Pravda, February 1, 1976.

78. Carey, op. cit., p. 591.

79. Bush, op. cit., pp. 15-16.

80. Diamond and Krueger, op. cit., p. 324.

81. Carey, op. cit., p. 587.

82. Bush, op. cit., pp. 16-18.

83. Carey, op. cit., p. 587.

84. Diamond and Krueger, op. cit., pp. 322-23.

85. Alec Nove, "Soviet Agriculture Under Brezhnev," Slavic Review, September 1970, pp. 386-89; Bush, op. cit., p. 27.

86. Bush, op. cit., pp. 20-21. See also by Bush, Radio Liberty Research Paper No. 29 (1968), pp. 25-26; Radio Liberty Dispatch, July 3, 1970 and August 31, 1971 (where he calculates the subsidy for 1960 through 1969). See Constance Krueger, "A Note on the Size of Subsidies on Soviet Government Purchases of Agricultural Products," The Association for Comparative Economic Studies Bulletin, Fall, 1974, pp. 63-68. (She estimates the subsidies by product for several years.)

87. Volin, op. cit., pp. 416-17.

88. See U.S. Central Intelligence Agency, USSR Agriculture Atlas, op. cit., p. 43.

89. Bush, Soviet Agriculture: Ten Years Under New Management, op. cit., p. 33.

90. See M. Gardner Clark, Agricultural Social Security and Rural Exodus in Italy (Ithaca, N.Y.: Cornell University, Center for International Studies, January 1977). For an excellent analysis of relatively minor Soviet problems see Norton Dodge, "Recruitment and the Quality of the Soviet Agricultural Labor Force," in The Soviet Rural Community, ed. James R. Miller (Urbana, Ill.: University of Illinois Press, 1971), pp. 180-213.

91. Narodnoe khoziaistvo SSSR v 1974 godu, p. 455.

92. David K. Shipler, "Soviet 'Factories' Press the Industrialization of Agriculture," New York Times, October 12, 1976, p. 58, describes a complex created five years ago when 13 collective farms combined their cattle herding operations into one unit. By 1974 there were 505 interfarm livestock enterprises. See Narodnoe khoziaistvo SSSR v 1974 godu, p. 434.

93. See "Chicken Farming--a Successful Branch of Soviet Agriculture," Radio Liberty Dispatch, July 1, 1976, pp. 1-2.

94. See Ekonomika sel'skogo khoziaistvo, no. 8 (1974): 57; no. 4 (1976): 22.

95. Robert C. Stuart, "Aspects of Soviet Rural Development," Agricultural Administration, no. 2 (1975): 170; Narodnoe khoziaistvo SSSR v 1974 godu, p. 430.

96. Bush, Soviet Agriculture, op. cit., pp. 37-38.

97. Ibid., p. 35; Paige Bryan, "Progress of Reform in the Kolkhoz Councils," Radio Liberty Dispatch, October 23, 1973.

98. See Stuart, op. cit., p. 165.

99. Pravda, August 5, 1964. Cited by Dimitry Pospielovsky, "The Link System in Soviet Agriculture," Soviet Studies, April 1970, p. 411. Beznariadnoe zveno can be translated as links that are "normless" or "result-oriented," or "nonscheduled," or "not tied to the specific job assignment."

100. See Pospielovsky, op. cit., pp. 421-25, for this and other comparative results.

101. Ibid., p. 426.

102. See Keith Bush, "The Implementation of the Soviet Economic Reform," Radio Liberty Research Paper, No. 36 (1970), pp. 30-34.

103. Hedrick Smith, The Russians (New York: Quadrangle, 1967), pp. 212-14. Yanov confirmed this thesis in a lecture and discussions at Cornell University in 1975. For a report of another successful Khudenko experiment in Kazakhstan see the New York Times, April 17, 1970.

104. Literaturnaia gazeta, no. 11 (1973): 10; Radio Liberty Dispatch, April 12, 1973.

105. Andreas Tenson, "Soviet Agriculture--Marxist Dogma or Economic Expediency," Radio Liberty Research, August 18, 1976, pp. 1-3.

106. Keith Bush, "Brezhnev and Kosygin in Soviet Economic Performance," Radio Liberty Dispatch, March 2, 1976, p. 7.

107. Ibid.

108. D. K. Willis, "Harvest Outlook--Soviet Smiles," Christian Science Monitor, October 27, 1976; Carey, op. cit., pp. 588-92.

109. F. Douglas Whitehouse and Joseph F. Havelka, "Comparisons of Farm Output in the US and USSR, 1950-1971," Joint Economic Committee, U.S. Congress, Soviet Prospects for the Seventies (Washington, D.C.: Government Printing Office, 1973), p. 362.

110. I. Kurtsev, Ekonomika sel'skogo khoziaistva, no. 1 (1975): 20; Karl-Eugen Wädekin, "Investment and the Goal of 'Industrialization' in Soviet Agriculture," Radio Liberty Research, April 13, 1976, p. 2.

111. Wädekin, "Investment and the Goal of 'Industrialization' in Soviet Agriculture," op. cit., p. 4.

112. Gottfried von Haberler, The Theory of International Trade (New York: Macmillan, 1950), p. 125; A. A. Alchian and W. R. Allen, Exchange and Production Theory in Use (Belmont, Calif.: Wadsworth, 1969), pp. 9, 203-04, 214. For more recent versions of comparative advantage see papers on "Comparative

Costs and Economic Development," American Economic Review, May 1964, especially pp. 426-30.

113. U.S. Central Intelligence Agency, Handbook of Economic Statistics 1976, op. cit., p. 131.

114. Ibid., pp. 131, 134.

115. The Economist, August 14, 1976, p. 71.

116. USSR, Tsentral'noe statisticheskoe upravlenie, Narodnoe khoziaistvo SSSR.

117. U.S. Central Intelligence Agency, The Soviet Grain Balance, 1960-73 (Washington, D.C.: CIA, September 1975), p. 2. The data refer to calendar years. Blanks indicate data not available.

118. See U.S. Department of Agriculture, Foreign Agricultural Circular--Grains (Washington, D.C.: USDA, October 27, 1976), p. 2.

CHAPTER

2

SOVIET AGRICULTURE: SUCCESS OR FAILURE?

Harry G. Shaffer

Western Sovietologists have long been more critical of Soviet agricultural performance than of the performance of any other sector of the Soviet economy. Soviet agriculture is ridiculed ("if the Sahara were to be collectivized, they would soon be importing sand"), is described as utterly inefficient ("just compare her agricultural output per acre or per farm worker with that of the United States"), is labeled an incredible "mess" and a "miserable failure" ("after more than fifty years, they still can't feed their people and must buy wheat from the West"), and is allegedly based on "communist policies [which] have been disastrously counterproductive."[1] And it is understandable that Soviet agriculture has been selected as the prime target for such attacks. After all, has not agricultural output in the USSR increased incomparably more slowly than industrial output, construction, or services? Has it not been in the area of agriculture that ambitious plans have often been underfulfilled, even in years when plans for other sectors were overfulfilled? Yet, one cannot help but wonder: if agriculture has indeed been such an abysmal failure, how could the Soviet Union have advanced from a largely underdeveloped country, with only 10 percent of its labor force engaged in industrial pursuits, and with most of its peasants poor and illiterate--how could it have advanced from such humble beginnings

The author wishes to express his gratitude to Roy D. Laird for furnishing on several occasions specific statistical information and source references requested, and to M. Gardner Clark, David M. Schoonover, and Victor Perlo for reading the manuscript and making valuable suggestions for improvement. It goes without saying that the author alone is responsible for any interpretations and conclusions.

to a nation second only to the United States in economic power, scientific accomplishments, and military preparedness? How, without the necessary underpinnings from the agricultural sector, could the Soviet Union have made the economic progress it did in so relatively short a period of time without the aid, assistance, and foreign capital investments that paved the road to industrialization in the United States and elsewhere around the globe? How could it have done this without the necessary assistance from the agricultural sector during decades of virtually total blockade by the industrially more advanced nations in the West and in spite of the setbacks brought on by a revolution, a civil war, and a devastating World War II? And its total output adjudged by strongly anti-Soviet Western analysts as having increased more than fourfold during the past 25 years alone, from one-third to more than three-fifths of that of the United States (see Table 2.1)--how could it have risen so rapidly to so high a level with so dismal an agricultural record? Surely the grain purchased in the West during the past few years could not account for all that! Could it be that Western analysts have been grossly underestimating Soviet achievements in agriculture as they underestimated Soviet scientific and military capabilities a generation ago? True, Soviet agricultural development has not nearly equaled (and surely was never intended to equal) the impressive progress in Soviet industry. Still, might it be that Soviet agricultural performance, with all its undeniable weaknesses and shortcomings, has not been as unsatisfactory as most Western observers would have it? At the risk of playing the devil's advocate, is it not evident that Soviet agriculture must have played a role, and a major and reasonably adequate role at that, in providing the basis for the Soviet Union's remarkable overall economic growth?

TABLE 2.1

Soviet Gross National Product, 1950-75

Year	In Billions of 1971 Dollars	As Percent of U.S. GNP
1950	164	33
1960	311	45
1970	551	54
1975	686*	61

*Approximate.

Source: Keith Bush, "Soviet Economic Growth: Past, Present, and Projected," Radio Liberty Research Supplement, September 12, 1975. Figures derived from materials published by the U.S. Central Intelligence Agency, the Council of Economic Advisers, and the Organization for Economic Cooperation and Development.

I will not argue that Soviet agriculture is optimally efficient. Obviously, it is not. Nor will I argue that performance could not be greatly improved. Obviously it can. But I will argue that the Soviet Union's agricultural sector has contributed substantially to the country's overall economic development; that it has made reasonable progress over the years; and that many of the problems with which it has been beset have been unavoidable, while others have been due to errors committed in the past (the Soviets' own self-criticism testifies to that), but that the solution to these problems surely does not lie in the dismantling of the Soviet system of collective and state farms and the return to privately owned and privately operated farmsteads.

WHAT PRICE COLLECTIVIZATION?

To begin with, Western indictments of Soviet agriculture point to the cost of enforced collectivization, high in terms not only of human suffering but also of diminished agricultural output. Indeed, grain output dropped somewhat during the era of collectivization in the late 1920s and early 1930s as peasants' resistance mounted. Really drastic, however, was the decrease in livestock as peasants and their families slaughtered and ate their animals rather than turning them over to the collectives. The situation was further aggravated by the relentless state collections of grain which, in the face of diminished grain production soon caused widespread famine in the countryside, with obvious effects on the availability of fodder and therefore on livestock herds. Not until the mid-1930s did the situation begin to improve (see Table 2.2).

At first glance it would appear that at least in these early years collectivization proved indeed a failure. However, the judgment cannot be based on what one thinks should have been accomplished. Instead, the criterion has to be the extent to which collectivization enabled or prevented agriculture from fulfilling the tasks assigned to it then. And among priorities, maximization of short-run agricultural output was least important.

For better or for worse, the Soviet leadership's first priority was rapid industrialization at all costs. If ever the Soviet Union was to become a major world power, it had to industrialize. But, more than that, Lenin had deemed industrialization essential for the very survival of the young socialist state in a hostile environment; and by early 1931 Stalin made his prophetic prediction: "We are fifty or a hundred years behind the advanced countries. . . . We must make good this distance in ten years . . . or we shall go under."[2] The Nazi invasion of Soviet territory was then ten years and four months off.

TABLE 2.2

Grain Harvest and Livestock Herds, 1928-35 and 1940

	Grain Harvest (in millions of tons)	Livestock (in millions of head)		
		Cattle	Pigs	Sheep and Goats
1928	73.3	70.5	26.0	146.7
1929	71.7	67.1	20.4	147.0
1930	83.5	52.5	13.6	108.8
1931	69.5	47.9	14.4	77.7
1932	69.6	40.7	11.6	52.1
1933	68.4	38.4	12.1	50.2
1934	67.6	42.4	17.4	51.9
1935	75.0	49.3	22.6	61.1
1940	95.6	54.5	27.5	91.6

Sources: Sotsialisticheskoe stroitel'stvo, 1936 (Moscow, 1936), pp. 342-43, 354; Yu, Moshkov, Zernovaya problema v gody sploshnoi kollektivizasii (Moscow, 1966), p. 226; USSR Agriculture (Moscow: Novosti Press Agency Publishing House, 1975), p. 65; Zhivotnovodstvo SSSR (Moscow, 1959), p. 23.

To industrialize with maximum speed, all efforts had to be put into the rapid development of the producers' goods sector. The tasks assigned to the agricultural sector reflected this priority. The first task was to furnish the necessary food and raw materials to the cities and, if possible, provide a surplus for exports needed to pay for imported machinery and equipment; in other words, maximization of off-farm, marketable agricultural products, and not of total farm output was tho goal. The second task was to supply food and raw materials at a minimum price; resources could not be set aside for the production of large quantities of industrial consumer goods to pay for agricultural products. The final task was to provide a part of the badly needed increase in the number of factory workers: only large-scale farming could make farm mechanization and with it the permanent release of farm labor feasible; and only large-scale farming would allow the replacement of the old system under which, in the off season, many peasants worked part time, with a new one under which fewer peasants could work collectively full time, thereby releasing additional farm workers for work in the cities during the better part of the year.

After the Revolution, land had been distributed to some 25 million families.[3] Agricultural output began to recover from the setbacks brought on by the war, the Revolution, and the civil war.* By 1928, it had once again reached prewar levels.[4] The peasants who continued to cultivate their land in fragmented small holdings, largely by antiquated methods, produced more and had more to eat. However, quantities of agricultural products offered for sale remained below prewar levels. By 1927 it had become evident that the peasants, acting as rational economic men, were not willing to sell enough grain at low official prices, preferring to hoard it or feed it to the livestock. Urbanization called for increasing off-farm consumption of food and agricultural raw materials; but by January 1928, state procurements were one-third lower than a year earlier.[5] Soon, bread had to be rationed in the cities. Obviously, the system of small-scale private farming failed to meet the exigencies of the times. Add to this that, ideologically, private farming was deemed undesirable and the solution was obvious. Collectivization would facilitate effective party-state control over production and sales; it alone could enable the agricultural sector to fulfill the tasks assigned to it during the era of rapid industrialization; and simultaneously it could help to "wean the peasants" from their love for private ownership and to provide for them a "training ground for socialism." (Even the 1961 Party Program still referred to the collective farms as "a school of communism for the peasantry.")[6]

In light of the goals to be accomplished, collectivization seemed the only way; in light of the peasants' refusal to collectivize voluntarily, some degree of coercion would appear to have been necessary. On the other hand, the excesses of Stalin and of the men under him were certainly uncalled for, inexcusable, and moreover counterproductive. Soviet sources may not be critical of Stalin's collectivization drive per se; they may refer to the drop in livestock herds as a "cost . . . inevitable in any drastic breakup of old social relations"; but even they cannot deny that "these losses might have been much smaller if . . . excessive haste had not been shown as collectivization was being effected."[7] Although it is always risky to draw parallels, it is probably safe to say that the experience of most of Eastern Europe, of China, and of Cuba seems to indicate that collectivization can be achieved without such extreme measures of coercion.

---

*After a hectic period of "War Communism" (1917-21), land and much of small-scale trading and handicrafts were temporarily returned to private enterprise. This New Economic Policy (NEP) represented, so to speak, one step backward to facilitate subsequently two, more rapid, steps forward toward socialism. It lasted from 1921 to the introduction of the First Five-Year Plan in 1928.

TABLE 2.3

Marketed Quantities of Grain, 1928-40, and Percentage of Total Agricultural Output Marketed, 1913, 1937, and 1939

I. Marketed Output of Grain (Annual average, millions of tons)

1928-32	1933-37	1939-40
18.2	27.5	32.1

II. Agricultural Output Marketed (in percent of total)

1913 (Russia)	1937	1939
20.3	28.8	36.2

Sources: I. Narodnoe khoziaistvo SSSR v 1961 godu (Moscow, 1962), p. 341; II. Arcadius Kahan, "The Collective Farm System in Russia: Some Aspects of Its Contribution to Soviet Economic Development," in Agriculture in Economic Development, ed. Carl Eicher and Laurence Witt (New York: McGraw-Hill, 1964), p. 259.

The Soviet Union's agricultural sector played the role designed for it during the pre-World War II era of industrialization. Irrespective of output, irrespective of what was left to meet the peasants' own needs, it was made to provide increasing quantities of grain and other agricultural resources for the cities and some for export (see Table 2.3), and it released substantial numbers of workers for the factories.* In this manner, the Soviet peasantry was made to foot the major part of the industrialization bill. Out of the peasants' deprivation came much of the necessary forced savings. But the goal was accomplished: in a mere dozen years, between 1928 and 1940 (an era during which the Western world languished under severe depression and massive unemployment) the Soviet Union was catapulted into a mighty industrial power capable of producing most of the

*Between 1926 and 1939, 19.7 million people moved from rural to urban areas; and between 1928 and 1940, the number of Soviet workers outside of agriculture increased from 9 to 28.2 million.[8]

industrial goods, including most of the weapons for their war machinery, so essential in the eventual defeat of the Nazi invaders. According to Soviet sources, industrial output increased almost sixfold during this 12-year period;[9] among the best known Western estimates, the most conservative grants a better than two-and-a-half-fold increase while some of the others come closer to the Soviet figures than to those of their Western colleagues.[10] Even as critical an expert observer of the Soviet agricultural scene as Naum Jasny could not deny that "the socialization drive in [Soviet] agriculture achieved to a large extent its major purpose of serving as a basis for the industrialization drive."[11] The use of Soviet agriculture as a means to increase substantially Soviet living standards had to await a later period.

THE REST OF THE STALIN ERA: YEARS OF WAR AND RECOVERY

By 1940, Soviet gross agricultural output was some 15 to 20 percent higher than the 1928 level;[12] and food production had risen by more than 63 percent, according to Western sources,[13] hence by considerably more than the increase in population estimated at roughly 26 percent (from some 150 to about 190 million Soviet citizens).[14] Then came the most devastating land invasion in world history that cost the Soviet Union some 20 million lives and 30 million injured. The Germans occupied much of the Soviet Union's most fertile lands, destroyed or plundered more than 70,000 villages, 98,000 collective farms, 1,876 state farms, and 2,890 machine tractor stations, carrying off scarce agricultural equipment and decimating the still ailing livestock herds in the process. Interestingly, as a sideline, the German leaders decided to maintain the kolkhoz* system in the occupied territories because they, as the Soviet leadership a dozen years earlier, considered it the best system not necessarily to maximize output but to "avoid, as far as possible, halts in production and interruptions in the delivery of agricultural products."[15]

By the end of the war in 1945, grain output was a mere 47.3 million metric tons, less than half of what it had been in 1940,[16] livestock herds had drastically declined (see Table 2.4), total gross agricultural output had dropped by almost one-half according to Soviet sources and by at least one-third according to some Western sources,[17] and food output was down by some 55 percent.[18]

---

*Collective farm.

TABLE 2.4

Number of Livestock, 1941 and 1945
(in millions of head)

	January 1, 1941	January 1, 1945	Decline (in percent)
Cattle	54.4	44.2	18.9
Hogs	27.5	8.8	68.0
Sheep	79.9	57.9	27.5
Goats	11.7	12.3	(5.1)*
Horses	21.0	9.9	52.9

*Increase.
Source: Zhivotnovodstvo SSSR (Moscow, 1959), p. 23.

Young and healthy labor to rebuild the war-torn economy was in short supply, a casualty of the years of fighting and of German extermination policies in prisoner-of-war camps; and the city called ever-increasing numbers of those who had returned from the war, and of those who came of age, from the farms to the factories. Inventories of farm equipment had dropped greatly during the war (see Table 2.5). Much of the equipment that was left was worn out and virtually unfit for use, and the small quantities of fertilizer that had been produced annually before the war were down by more than two-thirds from the prewar peak in 1939.* With agriculture in such a sad shape Stalin--warning of the impending encirclement of the Soviet Union by the "imperialists"--persisted in his policy of high priority for the producers' goods sector, imposing the cost of continuing rapid industrialization as before primarily on the peasants. In line with this policy, the state paid on the average about the same procurement prices for major food staples to collective farmers at the end of the 1940s as it had paid in 1937 (more for some items, less for others--see Table 2.6) while official retail prices of manufactured goods were allowed to increase more than two-and-a-half-fold.[20] And from 1946 to 1950, the state invested only 2.67 billion rubles in agriculture, as compared with over 15 billion rubles in heavy industry and almost

---

*Output of mineral fertilizer was 772,900 metric tons in 1939 and 253,100 tons in 1945.[19]

TABLE 2.5

Soviet Farm Equipment, 1940 and 1945

	1940	1945	Decline (in percent)
Tractors (in thousands)	531	397	25
Combines (in thousands)	182	148	19
Trucks (in thousands)	228	62	73
Combined draft power (tractors and horses, in millions of horsepower)	40.1	22.6	44

Source: Narodnoe khoziaistvo SSSR v 1963 godu (Moscow, 1964), pp. 378, 380.

TABLE 2.6

Average Procurement Prices, 1937 and 1950
(rubles per metric ton)

	1937	1950	1950 as Percent of 1937
Grain	88	91	103.4
Potatoes	53	40	75.5
Meat (live weight)	950	700	73.7
Milk*	240	300	125.0
Sugar beets	44	110	250.0

*Includes milk product converted to milk equivalents.

Source: Nancy Nimitz, "Soviet Agricultural Prices and Costs," Comparisons of the United States and Soviet Economies, Joint Economic Committee, U.S. Congress (Washington, D.C.: Government Printing Office, 1960), Pt. I, p. 251.

5 billion rubles in transportation.* It is in light of these conditions and of these policies that we must evaluate Soviet agricultural performance during the post-World War II Stalin years.

The Fourth Five-Year Plan (1946-50) called for an increase in agricultural output which, by the end of the period, would have had Soviet agriculture surpass the 1940 level by 27 percent. The plan was not fulfilled; but within this five-year period, Western and Soviet sources agree, agricultural output had once again reached the prewar, 1940 level[22]--quite an accomplishment in its own right when one considers the sad state of Soviet agriculture at the end of the war. By the time of Stalin's death in 1953, even livestock herds, so decimated during the war, had well surpassed their prewar (although not yet their precollectivization) numbers (see Table 2.7). But, even more important, the Soviet Union's agricultural sector once again fulfilled the primary task assigned to it: it was instrumental in the industrial sector's rapid recovery from the devastation of the war and it facilitated continuous high industrial growth rates thereafter. According to Soviet sources, industrial output was 73 percent above the prewar level in 1950 and about 150 percent above it in 1953;[23] more conservative Western estimates are considerably lower but still respectable, showing industrial output at 35 and 67 percent above its prewar level for 1950 and 1953 respectively.[24] The level of industrial output in 1945 was estimated by these Western sources at more than 30 percent below that of 1940, which would make the 1950 industrial output just about twice, and the 1953 output almost two-and-a-half times that of 1945. In the absence of massive importation from abroad, the food and raw materials simply had to come from the Soviet Union's own agricultural sector. This does not mean that that sector's contribution to industrial growth was optimal; but neither could it possibly have been terribly inadequate.

One can, of course, argue--and correctly so--that Soviet agriculture at the end of the Stalin era was badly neglected, underequipped, and inefficient in terms of output per person and per acre as compared with, for example, the United States. One can further argue that its contribution to overall Soviet economic growth was achieved largely by coercion and the threat of coercion, and by virtually confiscatory procurement prices that kept the purchasing power and living standards of peasants low, thus effectuating part of the forced savings necessary for speedy capital formation. All this is surely true.

---

*It should be pointed out that the collective farms themselves also invested some 3 billion rubles during that time and that state investment in the consumers' goods sector was even less than in agriculture--some 2 billion rubles in all.[21]

But must one not judge Soviet agricultural performance in terms of the tasks assigned to it? Increasing output and higher living standards for the inhabitants of both urban and rural areas did not become major goals prescribed for Soviet agriculture until the post-Stalin era.

TABLE 2.7

Livestock Herds, January 1, 1953

	In Millions of Head	As Percent of 1940
Cattle	56.6	103.9
Hogs	28.5	103.6
Sheep	94.3	113.3
Goats	15.6	133.3
Horses*	15.3	72.9

*As elsewhere in industrializing countries, horses became less and less important and finally largely vanished from much of the agricultural scene.

Source: Zhivotnovodstvo SSSR (Moscow, 1959), p. 23.

THE KHRUSHCHEV ERA: WINDS OF CHANGE IN SOVIET AGRICULTURE

With Stalin's death and the rise of Khrushchev to power, the era of ruthless exploitation of the peasantry came to an end. The development of the producers' goods sector remained one of the top priorities; but it was no longer the one and only goal to which all else had to be sacrificed. Thus, tribute was no longer to be exacted from the peasantry to pay for forced industrialization. On the contrary, with increases in agricultural outputs and in living standards for all among the goals to be achieved, the agricultural sector was henceforth to be given "all necessary" support. To this end, procurement prices were continuously increased, until, by 1962, they reached more than three times their 1952 level (see Table 2.8);*

*The era of rising farm prices under Khrushchev came to an end in 1963 with the further rise in procurement prices for cotton, sugar beets, and potatoes.

TABLE 2.8

Indexes for Average State Purchase Prices
for Selected Agricultural Products
(1952 = 100)

	1953	1954	1960	1962
Grains	236	739	717	845
Livestock	385	579	1,243	1,523
Cotton (unginned)	105	102	107[a]	--[b]
Potatoes	316	369	836[a]	--
All agricultural products	154	207	299	332

[a]1959.

[b]Data not available.

Sources: Central Statistical Administration, Sel'skoe khoziaistvo SSSR (Moscow, 1960), p. 17; S. G. Stolianov in Tsenoobrazovanie na prodmety potrebleniia, ed. Sh. Ia. Turetskii (Moscow, 1964), p. 115.

TABLE 2.9

Investment in Agriculture, 1946-64
(in comparable prices)

Year	Gross Investment in Agriculture[b]	Percent of Total Gross Investment	Gross State Investment in Agriculture[b]	Percent of Total Gross State Investment
1946-50[a]	1,158	15.2	535	7.7
1952	2,140	16.2	1,067	8.7
1953	2,153	15.5	985	7.7
1954	3,217	19.6	1,792	11.9
1955	4,365	23.5	2,265	13.8
1956	4,654	21.8	2,409	12.6
1957	4,886	20.5	2,703	12.5
1958	5,526	20.2	2,683	10.9
1959	6,021	19.5	2,495	9.1
1960	6,227	18.3	3,061	9.9
1961	6,882	19.2	3,727	11.4
1962	7,454	19.7	4,180	12.0
1963	8,213	20.3	4,797	13.0
1964	9,695	21.9	5,786	14.3

[a]Annual average.

[b]In millions of rubles.

Source: From Soviet sources in Lazar Volin, A Century of Russian Agriculture (Cambridge, Mass.: Harvard University Press, 1970), pp. 348-49.

guaranteed minimum incomes were gradually introduced into the collective farm sector; and the equalization of city and country living standards became a frequently reiterated, even if not completely attained, goal.* In line with the economic reforms in the industrial sector, a limited degree of decentralization of economic decision making was introduced into kolkhoz administration, investments in agriculture were greatly increased (see Table 2.9), collective farm chairmen were given somewhat greater flexibility in deciding on the use of investment funds, and the level of educational training and experience of top kolkhoz management improved significantly as the years went by. Almost from the outset of the Khrushchev era, the years of Stalin terror came to an end. By 1958, even the MTSs, former tools of effective state control over collective farms, were abolished and the farm machinery sold to the collective farms.† The degree of actual liberalization varied over the years, but the extremely oppressive atmosphere of the Stalin era had become a thing of the past.

It would go beyond the scope of this chapter to evaluate in detail the various agricultural experiments that were tried and the various schemes that were introduced by Khrushchev, a man whom an outstanding Western authority on Soviet agriculture labeled the "agrarian leader."[26] Some of these projects met with a certain measure of success; others did not. For instance, his corn program that greatly expanded the acreage under corn--from a former 4.5 million hectares to a peak of 37 million hectares in 1962--was judged "on balance . . . successful" by some Western observers,[27] although part of the area newly planted to corn soon proved ill-suited for corn production. The virgin lands campaign initiated in 1954

---

*The 1961 Party Program states that the "two basic, closely related tasks" of agriculture are: "(a) to build up an abundance of high-quality food products for the population and of raw materials for industry and (b) to effect the gradual transition of social relations in the Soviet countryside to communist relations and eliminate, in the main, the distinction between town and country."[25]

†MTSs were state-operated machine tractor stations which supplied collective farms with farm equipment and with necessary operators. Partly needed because of inadequate supplies of farm machinery and of workers capable of operating machinery and keeping it in repair, the MTSs also served the political purpose of facilitating control over collective farms. The mere threat of being the last to be furnished necessary equipment during harvesting time became a powerful weapon of control. And as state employees, the MTS personnel were more trustworthy than the more independent collective farm chairmen.

brought some 40 million hectares of virgin and long-fallow lands under the plow, mostly in Kazakhstan and Siberia, and had a highly favorable short-run impact on agricultural output, which, however, tapered off after the first few years (see Table 2.10). But in recent years, the virgin lands have again contributed significantly to the Soviet Union's grain output--some 27 percent of the total in the mid-1970s.[28] On the other hand, the virtual abolition of the grassland system of crop rotation (without enough fertilizer and cultivation equipment to make such an undertaking a halfway acceptable risk) proved costly in the long run in terms of soil deterioration and depletion of fertility that is not easy to restore.

TABLE 2.10

Estimated Production of Grain from the "New Lands," 1954-63

Year	Millions of Metric Tons
1954	4.5
1955	8.0
1956	25.0
1957	13.0
1958	23.0
1959	16.0
1960	18.0
1961	15.0
1962	17.0
1963	10.0

Source: U.S. Central Intelligence Agency, The Production of Grain in the U.S.S.R. (Washington, D.C.: CIA, October 1964), p. 17.

Many of Khrushchev's agricultural policies were obviously ill-conceived and failed to yield anticipated results. Moreover, ambitious plans for agricultural growth were often underfulfilled, especially after 1958. Yet, considerable progress was achieved in less than a dozen years, from Khrushchev's rise to power to his dismissal in the fall of 1964.* Grain output almost doubled between 1953 and 1964, allowing for gradually increasing grain exports and increasing shares of output for state grain procurements (see Table 2.11). Output of livestock products, likewise, almost doubled and livestock herds were built up substantially (see Tables 2.12 and 2.13). Total agricultural output, according to Western sources, rose by an annual average of 4.4 percent during the 1950s and 2.3 percent from 1960 to 1964[29]--a healthy 73 percent in all for the period from 1952 to 1964 (Table 2.14). Peasants' income increased almost two-and-a-half-fold during that period (Table 2.15). And although--for reasons discussed in detail elsewhere--industrial growth rates slowed down in the late 1950s and early 1960s,[30] they still averaged a respectable 7.6 percent from 1961 to 1965 (Table 2.16). Overall Soviet growth (GNP), finally, was computed by Western sources as averaging 5.2 percent for 1961-65, down somewhat from the 6.4 annual average for 1951-60, but still substantially higher than growth rates in the United States and in most of the rest of the world.[31] Once again, is it not obvious that such growth could not have been achieved without adequate agricultural performance, especially during a period when living standards were rising (see, for example, output of consumers' nondurables in Table 2.16) and when sizable net grain exports totaled millions of tons per year (see Table 2.11).

It can surely be argued that even greater success could have been achieved had it not been for certain mistakes and errors of judgment. Shifting some of the investments from the new lands in the East to more fertile lands in the West, loosening even further the reins of central control, continuing a grassland system of crop rotation--these and other alterations in policies might well have proven the better part of wisdom. Still, while justifiable criticism along these lines can surely be levied against the Khrushchev regime, substantial progress of Soviet agriculture during the Khrushchev years seems self-evident. But the fact remains that (perhaps in small part because of errors committed and of ill-conceived policies imposed on the agricultural sector, but surely primarily because of less favorable weather conditions) agricultural growth slowed down after 1958. The great agricultural setback of 1963 (see Tables 2.11 and 2.14) brought on by a severe drought, probably played a major role in Khrushchev's dismissal the following year and his replacement by the present Soviet leadership.

---

*Khrushchev became First Secretary of the Central Committee of the Communist Party of the Soviet Union on September 7, 1953, and Chairman of the USSR Council of Ministers--frequently referred to as "Premier" or "Prime Minister"--on March 27, 1958. On October 14, 1964, Brezhnev became the new First Secretary; and the following day, Kosygin became the new Soviet Premier.

TABLE 2.11

Grain Output, Procurements, Exports, and Imports, 1953-64

Year	Grain Output (million tons)	Grain Procurements (percent of total grain output)	Grain Exports (million tons)	Grain Imports (million tons)
1953	82.5	31.1	--	--
1954	85.6	39.5[a]	--	--
1955	103.7	--[b]	3.7	0.31
1956	125.0	--	3.2	0.59
1957	102.6	--	7.4	0.15
1958	134.7	42.0	5.1	0.78
1959	119.5	39.0	7.1	0.26
1960	125.6	37.2	6.8	0.24
1961	130.9	39.8	7.5	0.68
1962	140.1	40.4	7.8	0.05
1963	107.4	41.7	--	--
1964	152.0	44.9	--	--

[a]Average, 1954-58.

[b]Data not available.

Sources: Grain output from Soviet sources: for 1953-58 in Alec Nove, An Economic History of the USSR (London: Allen Lane, The Penguin Press, 1969), p. 334; for 1960-64 in U.S. Central Intelligence Agency, The Soviet Grain Balance, 1960-73 (Washington, D.C.: CIA, September 1975), p. 7; for 1959 from Narodnoe khoziaistvo SSSR v 1968 godu (Moscow, 1969), p. 349. Grain procurements from Jerzy F. Karcz, "Seven Years on the Farm: Retrospect and Prospects," Joint Economic Committee, U.S. Congress, New Directions in the Soviet Economy (Washington, D.C.: Government Printing Office, 1966), pt. II B, p. 411. Grain exports and imports from S. Kabysh, "The Crisis in Soviet Agriculture--I," Committee on the Judiciary, U.S. Senate, The Many Crises of the Soviet Economy (Washington, D.C.: Government Printing Office, 1964), p. 49.

TABLE 2.12

Index of Physical Output of Animal Products, 1953-64
(1950 = 100)

Year	Index
1953	119
1954	123
1955	137
1956	145
1957	160
1958	172
1959	185
1960	184
1961	200
1962	204
1963	199
1964	186

Note: Includes changes in inventories of livestock, weighted by 1953 prices. Amounts of grain, potatoes, and milk fed to livestock are deducted.

Source: Douglas B. Diamond, "Trends in Output, Inputs, and Factor Productivity in Soviet Agriculture," Joint Economic Committee, U.S. Congress, New Directions in The Soviet Economy (Washington, D.C.: Government Printing Office, 1966), Pt. II B,, p. 36.

TABLE 2.13

Livestock Herds, 1953 and 1965

	January 1, 1953 (in millions of head)	January 1, 1965 (in millions of head)	January 1, 1965 as Percent of January 1, 1953
Cattle	56.6	87.2	154
Hogs	28.5	52.8	185
Sheep	94.3	125.2	133

Sources: Zhivotnovodstvo SSSR (Moscow, 1959), p. 23; Narodnoe khoziaistvo SSSR v 1964 godu (Moscow, 1965), p. 251.

TABLE 2.14

Index of Agricultural Output, 1952-64

Year	Index (1952 = 100)	Annual Growth Rate (in percent)
1952	100	6.0
1953	105	5.4
1954	109	3.8
1955	125	13.8
1956	139	12.2
1957	145	3.1
1958	157	8.2
1959	154	-1.5
1960	155	.9
1961	168	8.1
1962	164	-2.6
1963	143	-12.8
1964	173	20.9

Source: Douglas B. Diamond and Constance B. Krueger, "Recent Developments in Output and Productivity in Soviet Agriculture," Joint Economic Committee, U.S. Congress, Soviet Economic Prospects for the Seventies (Washington, D.C.: Government Printing Office, 1973), p. 335.

TABLE 2.15

Average Annual Payment per Farm Worker, 1952-63

Year	On Collective Farms (in rubles)	On State Farms (in rubles)	Index of Real Wages on Collective Farms (1953 = 100)[a]
1952	184	--[b]	--
1957	331	--	182
1958	311	574	206
1959	320	575	194
1960	332	565	183
1961	379	639	224
1962	424	754	224
1963	432	760	232

[a]Wages for work on collective farms only; in-kind payments valued in state retail prices.

[b]Data not available.

Sources: Lazar Volin, A Century of Russian Agriculture (Cambridge, Mass.: Harvard University Press, 1970), pp. 420-25. Last column from Nancy Nimitz, Farm Employment in the Soviet Union, 1928-63 (Santa Monica, Calif.: The Rand Corp., November 1965), p. 97.

TABLE 2.16

Average Annual Rates of Growth of Civilian Industrial Output, 1951-65
(in percent)

	1951-55	1956-60	1961-65
Total industry	10.7	9.5	7.6
Industrial materials	10.4	9.3	7.0
Civilian machinery, including electronics	12.4	13.0	11.3
Nondurable consumer goods	10.0	6.9	4.8

Source: Joint Economic Committee, U.S. Congress, Soviet Economic Performance, 1966-67 (Washington, D.C.: Government Printing Office, May 1968), p. 19.

THE BREZHNEV-KOSYGIN ERA: A NEW ORIENTATION

By the mid-1960s, when Brezhnev and Kosygin took over the reins of the Soviet Party and government, the USSR had become an industrialized, highly advanced, mature socialist country. Although the further rapid development of the producers' goods sector was certainly not to be abandoned, it became possible, at long last, to give high priority also to the Soviet consumer.* Recent five-year plans and the speeches of Soviet leaders all reflect this consumer orientation, and Western sources readily acknowledge the trend.†

*Actually, the producers' goods sector was still allowed to grow somewhat more rapidly than the consumers' goods sector during the Eighth and Ninth Five-Year Plan era; and it is scheduled to grow more rapidly during the Tenth Five-Year Plan era, 1976-80. See Table 2.24.

†A long section of the Ninth Five-Year Plan (1971-75), for instance, deals with the "Further Rise of the Living Standard and the Cultural Level of the Soviet People." It contains provisions for very substantial increases in wages, pensions, and scholarship funds, for introducing allowances for children and abolishing income taxes for low-income families, for greatly increasing the supply of consumer goods and services, and so forth. But the Ninth Five-Year

"Since assuming power in 1965," wrote two U.S. Sovietologists in a paper submitted to the Joint Economic Committee of the U.S. Congress, "the Brezhnev regime has committed itself to a policy of raising consumer welfare, especially through improvements in the Soviet diet."[33] And a recent CIA publication comments: "The [Brezhnev-Kosygin] leadership has committed itself to major programs in the consumption area. The most striking evidence of the new policy is the plan to make more meat and consumer durables . . . available to the people."[34]

With the Soviet consumer in mind, the Brezhnev-Kosygin regime put greater emphasis on the development of agriculture than had ever been done before. Investments in agriculture increased at average annual rates of 7.5 percent during 1965-70 and 10 percent* during the 1971-75 plan period;[35] by the mid-1970s, they reached the highest level in Soviet history, both in absolute ruble amounts and as a percentage of total investments. They are scheduled to increase again during the present 1976-80 five-year plan period (see Table 2.17). More specifically, quantities of fertilizer delivered to agriculture increased rapidly from year to year, reaching in 1975 almost three and a half times their 1964 level (Table 2.18); inventories of agricultural equipment rose substantially over the years (Table 2.19); many millions of hectares of newly irrigated and newly drained land were put under cultivation, and land reclamation is scheduled to continue unabated.† Under a new collective farm charter adopted in 1969 (the first new one since 1935), collective farmers are assured of a minimum remuneration for their work and are entitled to the same social security benefits as city workers or workers on state farms.[37] Procurement prices paid to collective farms have been increased repeatedly during the Brezhnev-Kosygin era--grain prices, for instance, by 12 percent for most regions and by as much as 53 percent for some in March 1965,[38] and by about

---

Plan still emphasized that the producers' goods sector "remains the foundation of the country's economic might and of the further growth of the people's living standards," while the Tenth Five-Year Plan refers to the "raising of the people's material and cultural living standard" as the main task of the Tenth Five-Year Plan.[32]

*The 10 percent figure, given as the plan for 1971-75, was easily reached since actual investments in agriculture during that period surpassed the planned 129 billion rubles.

†4.5 million hectares of newly irrigated land and 4.4 million hectares of newly drained land were put under the plow in 1971-75; and 4 million hectares are to be irrigated and another 4.7 million to be drained during 1976-80.[36]

TABLE 2.17

Investment in Agriculture, 1966-80
(in billions of 1969 rubles)

	1966-70	1971-75	1976-80 (planned)
Total investment in economy	352	500	625[b]
Investment in agriculture[a]	82	131	172
Investment in agriculture as percent of total investment	23.3	26.2	27.5

[a]Includes both state and collective farm investment; collective farm investment amounts to roughly one-third of total investment in agriculture.

[b]The Tenth Five-Year Plan calls for an increase in total capital investments of "24 to 26 percent" over the 1971-75 level.

Sources: Twenty-fourth Congress of the Communist Party of the Soviet Union, Directives for the Five-Year Economic Development Plan of the USSR for 1971-1975 (Moscow, 1971), p. 16; Twenty-fifth Congress of the Communist Party of the Soviet Union, "Guidelines for Soviet Economic Development," Soviet Life, March 1976, p. 2; Karl-Eugen Wädekin, "A Survey of Soviet Agriculture in 1974," Radio Liberty Research Supplement, May 9, 1975, p. 17.

TABLE 2.18

Fertilizer Delivered to Agriculture, Selected Years

Year	Millions of Tons
1964	22
1972	54
1973	58
1974	64
1975	73
1980*	115

*Planned.

Note: Not all the fertilizer produced in the USSR is delivered to agriculture. Most of the rest is exported; some is lost in transit. Output of fertilizer was, for instance, 25.6 million tons in 1964, 90.2 million in 1975, and is scheduled to reach 143 million tons in 1980.

Sources: Lazar Volin, A Century of Russian Agriculture (Cambridge, Mass.: Harvard University Press, 1970), p. 364; USSR Agriculture (Moscow: Novosti Press Agency Publishing House, 1973), p. 109; USSR Agriculture (Moscow: Novosti Press Agency Publishing House, 1975), p. 109; Pravda, December 14, 1975 and February 1, 1976.

22 percent between 1965 and 1972-73.[39] Finally, the average annual income of collective farmers rose very significantly during the Eighth and respectably during the Ninth Five-Year Plan periods, and is scheduled for another substantial increase during 1976-80; and although their average income is still only two-thirds that of factory and office workers (141 rubles for these latter as compared with 94 rubles for collective farmers),[40] differential wage increases, past and planned, are bringing country and town incomes closer together (see Table 2.20).*

TABLE 2.19

Tractor, Combine, and Truck Inventories in Soviet Agriculture, Selected Years

	1965	1970	1973	1974
Tractors	1,613	1,977	2,180	2,300
Grain combines	520	623	670	670
Trucks	945	1,201	1,284	1,400

Note: Count taken at year's end, in thousands.
Sources: For 1974, Sel'skaya zhizn', March 23, 1975, p. 2; for other years, SSSR v tsifrakh v 1973 godu (Moscow, 1974), p. 123.

The efforts of the Brezhnev-Kosygin regime to increase agricultural output and the supply of agricultural products to the Soviet consumer have met with considerable success. In spite of several years of adverse weather conditions, worst of all in 1975, agricultural output rose by almost 37 percent during the ten-year period of the Eighth and Ninth Five-Year Plans (1966-75), grain output by about the same, meat and milk production by roughly twice that percentage (see Table 2.21), and livestock herds grew somewhat more slowly but rather consistently over the years (see Table 2.22 for

---

*N. K. Baibakov, Vice-Chairman of the USSR Council of Ministers and Chairman of the USSR Planning Committee, gave the plan figures for 1976 as 98 rubles per month for collective farmers and 150 for factory workers, which was to entail an increase of 5 percent and 2.7 percent respectively over 1974.

TABLE 2.20

Increase in Average Annual Income over Preceding Five-Year Period, 1966-80 (in percent)

	1966-70	1971-75	1976-80 (planned)
Factory and office workers	26	20	16-18
Collective farmers	72	25	24-27

Sources: Twenty-fourth Congress of the Communist Party of the Soviet Union, Directives for the Five-Year Economic Development Plan of the USSR for 1971-1975 (Moscow, 1971), p. 8; Twenty-fifth Congress of the Communist Party of the Soviet Union, "Guidelines for Soviet Economic Development," Soviet Life, March 1976, pp. 2-3; Pravda, December 14, 1975.

TABLE 2.21

Agricultural Output: Gross Product and Selected Commodities, Annual Averages, 1961-75 (in millions of tons unless otherwise stated)

	1961-65	1966-70	1971-75	Index 1971-75 (1966-70=100)	Index 1971-75 (1961-65=100)
Gross agricultural product (in billions of 1965 rubles)	66.5	80.5	92.0	113.0	136.8
Grain	130.5	167.6	180.2*	107.4	138.1
Meat	7.9	11.6	14.1	121.6	178.4
Milk	51.7	80.6	87.5	108.6	169.2
Cotton	5.0	6.1	7.7	126.2	154.0

*The latest Soviet figure for the 1975 grain output is 140 million tons. If a lower figure of 134 million tons were substituted, the average for the five-year period would drop to 179.0 million tons, and the relevant indexes in the next two columns to 106.8 and 137.2 respectively.

Sources: Pravda, April 5, 1966 and December 14, 1975; Twenty-third Congress of the CPSU, 1966 (Moscow, 1966), pp. 170-71; Narodnoe khoziaistvo SSSR v 1973 godu (Moscow, 1974), p. 343; Twenty-fourth Congress of the Communist Party of the Soviet Union, Directives for the Five-Year Economic Development Plan of the USSR for 1971-1975 (Moscow, 1971), p. 39; Twenty-fifth Congress of the Communist Party of the Soviet Union, "Guidelines for Soviet Economic Development," Soviet Life, March 1976, p. 2. Figure for 1975 grain output from Pravda, February 1, 1976.

cattle and hogs). According to CIA "adjusted" annual average data, the number of people working on farms dropped during this period from about 40 to less than 35 million, and the percentage of the labor force engaged in agriculture from around 35 percent (it used to be over 75 percent before the Revolution and still almost 42 percent in 1960) to 25.5 percent (see Table 2.23). Moreover, the sizable advance in agriculture was accompanied by an even more remarkable growth of the industrial sector: output of industrial consumers' goods more than doubled during these past ten years; and the output of producers' goods rose by a spectacular 122 percent (see Table 2.24). With population increasing no more than 9 percent[41] and with prices of consumer goods and services unchanged during the entire period,* the substantial rise in the take-home pay of city and farm workers discussed above (and, by the way, of students and pensioners as well) must obviously have added up to significant increases in living standards for the Soviet people.[42]

TABLE 2.22

Number of Cattle and Hogs, 1965-76
(in millions of head)

Year*	Cattle	Hogs
1965	87.17	52.84
1966	93.44	59.58
1967	97.11	58.03
1968	97.17	50.87
1969	95.74	49.05
1970	95.16	56.06
1971	99.22	67.48
1972	102.43	71.43
1973	104.01	66.59
1974	106.27	70.03
1975	109.10	72.30
1976	109.12	72.27

*January 1 of each year.

Sources: U.S. Central Intelligence Agency, Handbook of Economic Statistics 1975 (Washington, D.C.: CIA, 1975), pp. 130-31; idem, Handbook of Economic Statistics 1976 (Washington, D.C.: CIA, September 1976), pp. 132-33.

---

*Soviet claims of an unchanged consumer price index are supported by CIA studies.

TABLE 2.23

Labor Force, 1960 and 1965-75
(in millions of persons at midyear)

	Total Labor Force	Industrial Labor Force	Agricultural Labor Force	Agricultural Labor Force as Percent of Total Labor Force
1960	110.13	22.62	46.11	41.87
1965	116.48	27.45	40.18	34.50
1966	118.11	28.51	39.39	33.35
1967	119.86	29.45	38.66	32.25
1968	121.42	30.43	37.70	31.05
1969	123.56	31.16	37.88	30.66
1970	110.13	32.62	46.11	25.42
1971	127.64	32.03	36.94	28.94
1972	129.69	32.46	36.34	28.68
1973	131.60	32.88	35.74	27.16
1974	133.58	33.37	35.13	26.30
1975	136.12	34.03	34.61	25.43

Sources: U.S. Central Intelligence Agency, Handbook of Economic Statistics 1975 (Washington, D.C.: CIA, 1975), pp. 42-43; idem, Handbook of Economic Statistics 1976 (Washington, D.C.: CIA, September 1976), pp. 45-46.

TABLE 2.24

Index of Industrial Output, 1966-80

	1966-70 (1961-65 = 100)	1971-75 (1966-70 = 100)	1971-75[a] (1961-65 = 100)	1976-80[b] (1971-75 = 100)
Total industrial output	150	143	216.5	135-139
Producers' goods	151	147[c]	222.0	138-142
Consumers' goods	149	137	204.1	130-132

[a]Computed.

[b]Planned.

[c]Estimated on basis of total industrial output and of consumers' goods output.

Sources: Twenty-fourth Congress of the Communist Party of the Soviet Union, Directives for the Five-Year Economic Development Plan of the USSR for 1971-1975 (Moscow, 1971), p. 8; Pravda, December 14, 1975; Twenty-fifth Congress of the Communist Party of the Soviet Union, The Land of Soviets Today and Tomorrow (Moscow, Novosti Press Agency Publishing House, 1976), p. 30; Communist Party of the Soviet Union Central Committee for the Twenty-fifth Congress, Guidelines for the Development of the National Economy of the USSR for 1976-1980 (Moscow: Novosti Press Agency Publishing House, 1976), p. 27.

The growing quantities of food staples produced in the Soviet Union are even more impressive if viewed in relation to world and to U.S. output of these items. With less than 7 percent of the world's population, the Soviet people have, in recent years, seen their agricultural sector produce over 11 percent of the world's meat, about 15 percent of the world's grain, 23 percent of the world's milk, and 25 to 30 percent of the world's wheat. In comparison with the United States, the Soviet Union has long produced more milk, even on a per capita basis; in good years, it approaches equality in grain output; and over the past ten years, it has increased its total meat production from less than one-half to over three-fifths--and its combined meat and fish production from three-fifths to over four-fifths--that of the United States (see Table 2.25). How all this is supposed to add up to an "abysmal failure," as has so often been maintained in the West, is difficult to understand.

In their all-out endeavor to improve the diet of the Soviet people, the Soviet leadership has embarked on a course of decreasing grain exports, at least in terms of percent of total supply (see Table 2.26), and of importing regularly substantial quantities of grain from abroad.* The primary reason for this attempt to further increase grain supplies available for domestic use has been the growing need for livestock feed, brought on by the shift toward more meat consumption. The amount of grain used to feed livestock has more than doubled since the last pre-Brezhnev-Kosygin five-year plan, and has risen also as a percent of the expanding grain supply (see Table 2.27). A composite picture of grain supply and usage over the years tells the story even better (see Table 2.28). It shows clearly that the much maligned grain "shortage" in the USSR does not mean that the Soviet people would go hungry were it not for grain imports from the West. It confirms that the shortage, such as it is, is the result of the shift in diet that comes with greater affluence, since it takes between 7.5 and 14 pounds of oats per pound of weight to raise a calf to a mature animal.†

---

*The Soviets engage in particularly heavy purchases of grain in years following agricultural setbacks--for instance some 23 million tons in 1973 and 25 million in 1976.[43]

†P. Esaulov, writing in Voprosy ekonomiki, no. 5 (1974), p. 63, gives the figure as 7.5 to 8 feed units (in the USSR, 1 feed unit equals 1 kilogram of feed grain) per kilogram of meat if it takes a year and a half to raise the animal to 400 kilograms and 10 to 12 feed units if it takes two years. In the United States, one must reportedly feed a steer 14 pounds of grain for each pound of meat produced; and 6 pounds of feed are needed to produce one pound of poultry.[44]

TABLE 2.25

Output of Grain, Wheat, Meat, Fish, and Milk: World, United States, and Soviet Union, 1965-75
(millions of metric tons)

	1965	1966	1967	1968	1969	1970	1971	1972	1973	1974	1975
Grain production											
World	986	1,060	1,110	1,140	1,150	1,160	1,250	1,220	1,320	1,290	1,410
U.S.	183	184	208	203	206	187	238	228	237	204	248
USSR	121	171	148	170	162	187	181	168	223	196	140
USSR, percent of world	12.3	16.1	13.3	14.9	14.1	16.1	14.7	13.8	16.9	15.2	9.9
USSR, percent of U.S.	66.1	92.9	71.2	83.7	78.6	100.0	76.1	73.7	94.1	96.1	56.5
Wheat production											
World	262	303	292	324	306	308	341	340	368	352	344
U.S.	36	36	41	42	39	37	44	42	46	49	58
USSR	60	101	77	93	80	100	99	86	110	84	66
USSR, percent of world	22.9	33.3	26.4	28.7	26.1	32.5	29.0	25.3	29.9	23.9	19.2
USSR, percent of U.S.	166.7	280.6	187.8	221.4	205.1	270.3	225.0	204.8	239.1	171.4	113.8
Meat production[a]											
World	73.3	75.8	78.6	81.0	82.1	84.4	86.8	87.9	88.2	92.8	96.4
U.S.	14.3	14.8	15.5	16.0	16.0	16.3	17.1	16.7	15.8	17.0	16.6
USSR	8.5	9.2	9.9	10.0	10.0	10.3	11.1	11.4	11.2	12.0	12.6
USSR, percent of world	11.6	12.1	12.6	12.3	12.2	12.2	12.8	13.0	12.7	12.9	13.1
USSR, percent of U.S.	59.4	62.2	63.9	62.5	62.5	63.2	64.9	68.3	70.9	70.6	75.9
Poultry production											
World	n.a.	n.a.	n.a.	n.a.	n.a.	17.7	18.2	19.6	20.1	20.1	n.a.
U.S.	4.9	5.3	5.6	5.5	5.8	6.2	6.3	6.6	6.5	6.5	6.5[c]
USSR	0.7	0.7	0.8	0.8	0.9	1.1	1.2	1.2	1.3	1.4	1.5[c]
USSR, percent of world	n.a.	n.a.	n.a.	n.a.	n.a.	6.2	6.6	6.1	6.5	7.0	n.a.
USSR, percent of U.S.	14.3	13.2	14.3	14.5	15.5	17.7	18.2	18.2	20.0	21.5	23.1[c]
"Other" meats,[b] including edible offal and lard											
World	n.a.	n.a.	n.a.	n.a.	n.a.	10.5	10.8	11.4	11.7	12.4	n.a.
U.S.	1.8	1.6	1.8	2.0	2.0	2.0	2.0	1.8	1.7	1.8	1.8[c]
USSR	0.8[c]	0.8[c]	0.8[c]	0.8[c]	0.9[c]	0.9	1.0	1.0	1.0	1.1	1.1[c]
USSR, percent of world	n.a.	n.a.	n.a.	n.a.	n.a.	0.9	0.9	0.9	0.9	0.9	n.a.
USSR, percent of U.S.	44.4	50.0	44.4	40.0	45.0	45.0	50.0	55.6	58.9	61.1	61.1[c]

Total meat production, including poultry, edible offal, and lard											
World	n.a.	n.a.	n.a.	n.a.	n.a.	112.6	115.8	118.9	120.0	125.3	n.a.
U.S.	21.0	21.7	22.9	23.5	23.8	24.5	25.4	25.1	24.0	25.3	24.9
USSR	10.0	10.7	11.5	11.6	11.8	12.3	13.3	13.6	13.5	14.5	15.2
USSR, percent of world	n.a.	n.a.	n.a.	n.a.	n.a.	10.9	11.2	11.4	11.3	11.6	n.a.
USSR, percent of U.S.	47.6	49.3	50.2	49.4	49.6	50.6	52.4	54.2	56.3	57.3	61.0
Fish catch (live weight)											
World	53.2	57.3	60.4	63.9	62.7	70.0	70.2	65.5	65.7	69.8	n.a.
U.S.	2.7	2.5	2.4	2.5	2.5	2.8	2.8	2.7	2.7	2.7	n.a.
USSR	5.8	6.1	6.5	6.8	7.1	7.8	7.8	8.2	9.0	9.6	10.3
USSR, percent of world	10.9	10.5	10.7	10.6	11.3	11.1	11.1	12.5	13.7	13.8	n.a.
USSR, percent of U.S.	214.8	244.0	270.8	272.0	284.0	278.6	278.6	303.7	333.3	355.6	n.a.
Total meat production, including poultry, edible offal, and lard plus fish catch (dressed weight)[d]											
World	n.a.	n.a.	n.a.	n.a.	n.a.	161.6	164.9	164.8	166.0	174.2	n.a.
U.S.	22.9	23.5	24.6	25.3	25.6	26.5	27.4	27.0	25.9	27.2	26.8[e]
USSR	14.1	15.0	16.1	16.4	16.8	17.8	18.8	19.3	19.8	21.2	22.4[e]
USSR, percent of world	n.a.	n.a.	n.a.	n.a.	n.a.	11.0	11.3	11.7	11.9	12.2	n.a.
USSR, percent of U.S.	61.6	63.8	65.4	63.6	65.6	67.2	68.2	71.5	76.4	77.9	83.2[e]
Milk production											
World[f]	342	348	355	363	362	362	365	372	378	385	386
U.S.	56.3	54.4	53.9	53.2	52.8	53.3	53.8	54.4	52.4	52.4	52.4
USSR	76.2	76.0	80.0	82.3	81.5	83.0	83.2	83.2	88.3	91.8	90.8
USSR, percent of world	21.2	21.8	22.5	22.7	22.5	22.9	22.8	22.4	23.4	23.8	23.5
USSR, percent of U.S.	129.0	139.7	148.4	155.0	151.5	155.7	154.6	152.9	168.5	175.2	173.3

(continued)

(TABLE 2.25 continued)

[a]Carcass weight, bone in basis, for beef, veal, pork, mutton, lamb, and goat. Excludes horsemeat, rabbit, poultry, game, edible offal, and lard.

[b]Includes horsemeat, rabbit, poultry, game, edible offal, and lard.

[c]Estimated.

[d]Dressed weight, comparable to carcass weight in meat, was obtained by applying a coefficient of 0.7 to live weight figures.

[e]Includes preliminary estimates of poultry, offal, "other meats," lard production, and, in the case of the United States, also of fish catch.

[f]In 1975-76 figures for world milk production for the 1960s and the 1970s were revised downward by 30 to 34 million tons annually. (Compare the Handbook of Economic Statistics 1975, p. 133 with the Handbook of Economic Statics 1976, p. 135.) Since figures for 1966-69 were not included in the 1976 edition, nor in the FOA Production Yearbooks, these have been estimated at 32 million tons below the previously published figures.

Note: n.a. = not available.

Sources: Compiled and computed from U.S. Central Intelligence Agency, Handbook of Economic Statistics 1975 (Washington, D.C.: CIA, 1975), pp. 123, 124, 129, 132, 133; idem, Handbook of Economic Statistics 1976 (Washington, D.C.: CIA, September 1976), pp. 125, 126, 131, 134, 135; U.S. Department of Agriculture, The Agricultural Situation in the Soviet Union, Review of 1975 and Outlook for 1976, Foreign Agricultural Economic Report No. 118 (Washington, D.C., April 1976), p. 11; Food and Agricultural Administration of the United Nations, FOA Production Yearbook, vol. 23 (Rome, 1969), pp. 354, 363; idem, vol. 25 (1971), pp. 362, 390; idem, vol. 26 (1972), pp. 200, 206-09; idem, vol. 28 (1974), pp. 218, 220, 221, 223, 227, 228. The coefficient used in note "d" was suggested in U.S. Department of Agriculture, Conversion Factors and Weights and Measures for Agricultural Commodities and Their Products, USDA/ERS Statistical Bulletin No. 362 (Washington, D.C., June 1965), p. 18.

TABLE 2.26

Soviet Grain Exports, Selected Years

	1960	1965	1970	1973	1976*
In millions of tons	6.9	5.3	6.8	6.2	0
In percent of total supply	5.5	4.1	3.6	2.5	0

*Preliminary estimate.

Sources: U.S. Central Intelligence Agency, The Soviet Grain Balance, 1960-73 (Washington, D.C.: CIA, September 1975), p. 14. Preliminary estimate in David W. Carey, "Soviet Agriculture: Recent Performance and Future Plans," Joint Economic Committee, U.S. Congress, Soviet Economy in a New Perspective (Washington, D.C.: Government Printing Office, 1976), p. 582.

TABLE 2.27

Grain Used for Livestock Feed, 1960-73

Year	Millions of Tons	Percent of Total Supply
1960	43.7	34.5
1961	48.8	37.1
1962	50.3	35.7
1963	42.6	38.3
1964	37.7	23.4
1965	53.8	42.0
1966	57.9	32.2
1967	61.6	40.8
1968	66.1	38.5
1969	74.9	45.8
1970	87.9	46.3
1971	93.4	50.4
1972	92.8	50.5
1973	99.2	40.2

Source: U.S. Central Intelligence Agency, The Soviet Grain Balance, 1960-73 (Washington, D.C.: CIA, September 1975), p. 14.

TABLE 2.28

Supply and Consumption of Grain, 1960-73
(in thousands of tons)

	1960	1961	1962	1963	1964	1965	1966
Supply	126,584.0	131,648.0	140,719.0	111,171.0	161,149.0	128,209.0	179,642.0
Domestic production	125,561.0	130,887.0	140,121.0	107,416.0	151,963.0	121,061.0	171,018.0
Imports	1,023.0	761.0	608.0	3,755.0	9,186.0	7,148.0	8,624.0
Consumption	124,994.4	133,693.7	138,172.8	127,306.8	119,783.7	137,921.3	143,390.8
Seed	24,152.4	25,845.2	27,832.8	27,956.8	27,191.7	26,399.3	25,767.8
Food	48,087.0	49,133.0	49,636.0	48,047.0	48,534.0	50,317.0	53,311.0
Flour	45,530.0	46,160.0	46,180.0	44,480.0	43,950.0	47,070.0	49,810.0
Groats	2,557.0	2,973.0	3,456.0	3,567.0	4,584.0	3,247.0	3,501.0
Industry	2,188.0	2,039.0	2,152.0	1,989.0	2,396.0	2,116.0	1,963.0
Alcohol	1,360.0	1,079.0	1,168.0	1,004.0	1,280.0	829.0	655.0
Beer	625.0	670.0	700.0	702.0	708.0	792.0	859.0
Starch and syrup	203.0	290.0	284.0	283.0	408.0	495.0	449.0
Exports	6,887.0	7,856.5	8,262.0	6,704.0	4,002.0	5,259.0	4,449.0
Livestock feed	43,680.0	48,820.0	50,290.0	42,610.0	37,660.0	53,830.0	57,900.0

	1967	1968	1969	1970	1971	1972	1973
Supply	150,825.0	171,651.0	163,660.0	189,745.0	185,243.0	183,940.0	246,987.0
Domestic production	147,749.0	169,291.0	162,255.0	186,795.0	181,175.0	168,238.0	222,530.0
Imports	3,076.0	2,360.0	1,405.0	2,950.0	4,068.0	15,702.0	24,457.0
Consumption	150,213.8	155,762.9	165,468.5	179,601.8	190,676.9	186,181.2	196,263.8
Seed	25,777.8	26,913.9	25,344.5	24,593.8	26,593.8	25,525.2	27,949.8
Food	53,729.0	54,000.0	54,242.0	57,480.0	58,560.0	59,370.0	60,030.0
Flour	49,930.0	50,000.0	50,040.0	52,980.0	54,060.0	55,270.0	55,430.0
Groats	3,799.0	4,000.0	4,202.0	4,500.0	4,500.0	4,100.0	4,600.0
Industry	2,269.0	2,378.0	2,497.0	2,780.0	2,898.0	2,992.0	3,080.0
Alcohol	879.0	898.0	987.0	1,148.0	1,204.0	1,242.0	1,280.0
Beer	903.0	958.0	993.0	1,048.0	1,100.0	1,150.0	1,200.0
Starch and syrup	487.0	522.0	517.0	584.0	594.0	600.0	600.0
Exports	6,868.0	6,351.0	8,525.0	6,848.0	9,745.0	5,464.0	6,204.0
Livestock feed	61,570.0	66,120.0	74,860.0	87,900.0	93,450.0	92,830.0	99,200.0

Source: U.S. Central Intelligence Agency, The Soviet Grain Balance, 1960-73 (Washington, D.C.: CIA, September 1975), p. 24.

While Western Sovietologists generally accept Soviet statistics on livestock herds, meat and milk production, and so forth, it should be pointed out that the same cannot be said for grain. Official Soviet figures for grain output are "adjusted" downward to account for what Western observers assert to be excess moisture, foreign matter, and loss in storage and transportation. One recent study, for instance, adjusted the official Soviet figure for average annual grain production for 1966-70 down from 167.5 million to 142 million metric tons, an adjustment of 15.2 percent; another did even better and adjusted the figure to 134.9 million tons, down 19.5 percent.[45] In years with considerable precipitation at harvest time, a northern country would obviously have excess moisture in its grain; but I have nowhere found any evidence that such substantial adjustments are called for as a matter of course on the basis of any proven differences between the moisture and waste content in Soviet and U.S. grain. Recently, a representative of the U.S. Department of Agriculture commented, "I believe excessive adjustments were made in the past, but we lack specific net weight data against which to evaluate adjustments."[46] As to losses in storage and transportation, a CIA study tested loss rates of from zero to 10 percent for "reasonableness" by computing what stock accumulations (an unknown quantity in the West) would have been had various assumed loss rates prevailed. That particular study then reached the conclusion that loss rates of 2 to 4 percent in storage and transportation would appear to be "reasonable."[47] While, under the circumstances, I find it rather difficult to accept such "adjustments" at face value, it should be pointed out that, in any case, they would affect only total output figures and not growth rates. Assuming that such losses, if indeed they exist, have remained fairly stable over the years, or at least over five-year plan periods, the relative growth of Soviet grain output would not have been altered; and growth rates of total grain supplies would actually have been affected favorably in recent years since Soviet domestic grain output has recently been augmented by the importation of allegedly superior Western grain that needs no downward adjustments.

Now that full credit has been given where credit has long been overdue, the fact still remains that Soviet agriculture is far behind U.S. agriculture in output per hour of labor time, per acre of land, and, in general, per unit of input. Although there is disagreement on the extent of the difference, the Soviets do not deny that they are lagging behind on all these counts and that, therefore, the real costs per unit of agricultural output produced are considerably higher in the Soviet Union than in the United States. How does one account for such lingering differences in agricultural performance?

To begin with, soil and climatic conditions in the Soviet Union are much worse than in the United States, "about 58 percent worse," according to one Soviet source.[48] Located as a whole much further north than the productive areas in the United States, the richest Soviet farmlands in the Ukraine are as far north as northern Montana and North Dakota. Kiev, Kharkov, and Karaganda are farther north than the northern border of the United States; and the southernmost tip of the Soviet Union is about as far south as San Francisco. The Baltic region, Byelorussia, and the center of the Russian Republic are often plagued by excessive moisture; and well over half of the plowed lands including the country's major grain-producing regions in the Ukraine, along the Volga, and in Kazakhstan suffer from inadequate precipitation. Moreover, high up north where rainfall is more reliable, soils are poor and growing seasons very short; the naturally fertile regions, on the other hand, are subject to recurrent droughts which generally occur during one or two years out of every five-year plan period. Severe droughts very frequently strike vast areas, sometimes all the way from the Altai to the Carpathians. That these natural drawbacks would stand in the way of more effective agricultural performance under any political, social, and economic system is obvious. Most Western analysts peripherally acknowledge them; but they usually fail to give them due weight in their comparisons of Soviet and U.S. agricultural performance. (Assume for a moment that Castro were to claim that Cuba's ability to produce more sugar per acre than the continental United States was due to Cuba's socialist system of agriculture. Would U.S. analysts not rightly ridicule such an assertion and, in their replies, put primary emphasis on differences in climate and soil?)

Much of the difference between Soviet and U.S. agricultural performance is explainable by the head start of the United States: American agriculture was mechanized, the countryside was electrified, and the fertility of the soil was restored or raised by the widespread application of mineral fertilizer decades before the Soviet Union effectively embarked on a similar course. In spite of all recent efforts, such advantage can be overcome but gradually. Even in the 1970s, the United States has, per unit of land, four times as many trucks as the Soviet Union, three and a half times as many tractors (the difference in horsepower is somewhat less since the Soviet tractors on the average are larger), and two and a half times as many grain combines. The difference per 1,000 farm workers is even more overwhelming: the United States has more than 17 times as many tractors per 1,000 farm workers as the USSR and almost 20 times as many trucks (see Table 2.29). But the gap has been narrowing constantly. The number of tractors, for instance, increased by almost 7 percent in the Soviet Union between 1971 and

TABLE 2.29

Inventories of Basic Farm Equipment: United States and Soviet Union, 1972

Equipment	United States	Soviet Union	United States as Percent of the Soviet Union
Tractors (in thousands)	4,387[a]	2,112	207.72
Total horsepower (in millions)	212.0	124.3	170.56
Tractors per 10,000 hectares of cropland	333	94	354.26
Total horsepower per 10,000 hectares of cropland	15,800	5,500	287.27
Tractors per 1,000 farm workers	1,002	58	1,724.14
Grain combines (in thousands)	703	656	107.16
Grain combines per 10,000 hectares of cropland	147	57	258.95
Trucks (in thousands)	2,915	1,232	236.61
Trucks per 10,000 hectares of cropland	221	55	401.82
Trucks per 1,000 farm workers[b]	665	34	1,955.89

[a]Not counting garden tractors, which add about another 850,000.

[b]1970.

Note: Tractors per 1,000 farm workers estimated by applying the same multiples to tractors per 10,000 hectares that are necessary to convert trucks per 10,000 hectares to trucks per 1,000 workers.

Sources: Except for last two rows, computed from U.S. Central Intelligence Agency, USSR Agriculture Atlas (Washington, D.C.: CIA, December 1974), p. 27. Trucks per 1,000 farm workers from F. Douglas Whitehouse and Joseph F. Havelka, "Comparison of Farm Output in the US and USSR," Joint Economic Committee, U.S. Congress, Soviet Economic Prospects for the Seventies (Washington, D.C.: Government Printing Office, 1973), p. 354. Garden tractor statistics from U.S. Department of Agriculture, Agricultural Statistics 1973 (Washington, D.C.: USDA, 1973), p. 5.

1973 and decreased by about 4 percent in the United States during the same period.[49] In the output of the mineral fertilizer, the Soviet Union was far behind the United States some 15 years ago. Between 1960 and 1973, however, the USSR quintupled its output of fertilizer.[50] (See also Table 2.18.) Although, according to Western sources, the quality of Soviet fertilizer is some 25 percent lower in terms of nutrient content than U.S. fertilizer,[51] the Soviet Union has by now surpassed the United States even in mineral fertilizer nutrients produced.[52] However, because of the much larger crop area to be covered (close to twice that of the United States), because of Soviet export commitments, and because of much heavier Soviet losses in transit (caused largely by less adequate transportation and storage facilities), the quantity per acre of fertilizer nutrient applied to crops is still almost twice as high in the United States as in the Soviet Union (it was 230 pounds and 111 pounds respectively in 1971, according to one Western source).[53]

Under the circumstances, it can come as no surprise that factor productivity in the Soviet Union's agricultural sector is still relatively low. During the past ten years, agricultural output increased by about 37 percent (albeit more slowly during the last than during the preceding five-year period), while labor inputs decreased by some 24 percent.* Hence, labor productivity rose by about 80 percent. With the land area devoted to agriculture unchanged or slightly expanding, production per unit of land has risen more gradually; and one Western source computed what appears to be a rather low estimate of a combined land-labor productivity index (with land weighted at 21 and at 31 percent) as on the average about 11 percent higher from 1965-71 in each case than during the first half of the 1960s[54]--an index that must have decreased somewhat in 1972, risen substantially in 1973, fallen back again in 1974-75, in the wake of agricultural setbacks, and risen considerably once more in the bumper crop year of 1976. Finally, as to capital productivity, the veritable surge in capital investments in agriculture could be expected to result in diminishing returns; but input-output ratios have certainly risen by more than was anticipated, and land-labor productivity did not benefit as much from the additional investments as had been hoped for.

To some extent, the less than satisfactory growth in factor productivity can be accounted for by natural causes such as the extreme drought of 1975. To some extent, it was undoubtedly the result of conscious decisions by the Soviet leadership to promote

---

*See Tables 2.21 and 2.23. Note that the increase in agricultural output is figured as 1971-75/1961/65 and the decline in the labor force as 1974/1965.

agriculture, even at the cost of investments that yield relatively low returns; to a considerable extent, surely, it was brought on by faulty judgment, unwise investment decisions, ill-advised policies, and plain inefficiency--and the Soviets are the first to admit it. A recent editorial in Izvestia, for instance, criticized the "wastefulness and carelessness that still survive in some quarters" of the agricultural sector;[55] Brezhnev, in a recent talk, referred to the "many unresolved problems that have been accumulating in the area of agriculture, some of them unavoidable, but others due to mistakes committed in the past," and he mentioned among current problems that need to be corrected "large losses incurring in the transportation and storage" of fertilizer.[56] And one can find articles critical of many aspects of Soviet agriculture, including the quality of farm equipment, the shortage of spare parts, or even the size of some of the state farms which, under Khrushchev, became so large as to be "unmanageable or poorly manageable."[57] Such willingness to recognize and analyze shortcomings has always been the first step toward corrective action; on such willingness lies much hope for the future.

SUMMARY AND CONCLUSIONS

In spite of undeniable shortcomings, Soviet agriculture has largely fulfilled the major functions assigned to it. During the Stalin era, with the constant drive for rapid industrialization, it provided much needed marketable supplies of food and raw materials for the growing urban population and for industry, while at the same time freeing substantial contingents of agricultural labor for industrial employment. Since then, it has, in addition, played the major role in greatly improving the diet and in substantially raising the living standards of Soviet citizens in towns and in the countryside alike. Its growth rates may not have been as high as hoped for; but they have been impressive just the same, particularly if one compares them with the pre-Revolutionary era from the mid-nineteenth century to the outbreak of World War I, when they averaged not more that 1 percent per year according to Soviet sources,[58] and 1.7 percent ("a figure only a fraction above the rate of population increase"), according to a recent Western estimate.[59] In an excellent study in which he uses almost exclusively Western data, even for the computation of Soviet growth rates, Charles K. Wilber compared the Soviet Union's agricultural growth with that of other nonsocialist countries during their corresponding period of development.[60] Whether in average or per capita growth rates, or in growth rates of output per worker, the Soviet Union compared very favorably with the others, even more so when the war and immediate postwar years, 1941-48, are omitted (see Table 2.30). Growth per unit of land is somewhat less impressive, but still respectable, especially for grain (the Soviets claim to have doubled average yield, from 7.7 to 15.4 centners per hectare, between the early 1920s and the early 1970s)[61] and for potatoes, vegetables, and industrial crops (see Table 2.31). In another interesting study, the United Nations compared the economic progress of the world, Africa, North America, South America, Asia, Europe, Oceania, and the USSR for the period 1952-70. The figures show the Soviet Union leading all others in growth of agricultural output and of food production, both in total volume and per capita (see Table 2.32). (The UN figures, by the way, do not seem too dissimilar from Western computations. Diamond and Krueger, for instance, computed an index of Soviet net agricultural production, 1950-72, from which 1970 can be calculated as 225 percent of 1952.)[62] Finally, during much of the first half of the 1970s, unusually unfavorable weather conditions adversely affected Soviet agriculture but did not prevent it from achieving what under the circumstances must be called reasonable progress during the Ninth Five-Year Plan period (1971-75). Today, the Soviet Union leads the world in the production of many agricultural products--including wheat, potatoes, sugar, milk, and butter--and holds second place in the world in the production of many others. Would it not be difficult to assess all this as a "miserable failure"?

The highly respectable performance of Soviet agriculture does, of course, not mean that it is highly productive. In yield per acre, in output per worker, perhaps even in marginal returns to capital, the Soviet Union lags far behind the United States. These undeniable facts are taken by many Western Sovietologists as prima facie evidence that it is largely (even if not exclusively) the Soviet system of socialized and collectivized agriculture that is responsible for this lag; and they generally propose as a solution that the USSR "disband the vast majority of the collective and state farms . . . and encourage intensified individual farming."[63] I feel that evidence does not warrant such a conclusion.

To begin with, a mere unadjusted comparison of Soviet and U.S. agricultural performance is of but limited value anyhow since allowance has to be made for different climatic and soil conditions, different backgrounds of farmers, different degrees of capital endowment, and so on. It would seem that these dissimilarities alone might well be sufficient to explain, for instance, most if not all of the difference in the wheat yield which, in 1971-72, was 15 centners per hectare in the USSR (1 centner equals 100 kilograms)[64] as compared with 22 centners for the United States. Those who would argue that the institution of private ownership of U.S. farms deserves the lion's share of the credit might find it difficult to explain why, then, in 1971-72 (a fairly normal year, weatherwise) the wheat yield in Communist Hungary was 30.80 and in Communist East Germany 39.5 centners per hectare--almost 80 percent more than in the United States; or why in that same year the wheat yield per hectare in Poland, where 80 percent or more of the land is in private hands, was 25.8 centners and in neighboring Czechoslovakia with its socialized agriculture 34.4 centners ?[65]

TABLE 2.30

Average Annual Growth Rates of Agricultural Output
(in percent)

Countries	Total	Per Capita	Per Worker
Soviet Union			
1928-1964/65	2.3	1.1	3.1[a]
1928-40 and 1948-64/65[b]	3.0	1.5	4.1
United States			
1870-1900	3.3	1.2	--[c]
1910-40	1.0	-0.2	1.7
1928-59	1.6	0.3	3.4
Great Britain			
1760-90	0.6	-0.3	--
1801/11-1831/41	1.2	-0.5	0.2
Germany			
1882-1909/13	1.8	0.6	0.6
France			
1830/34-1860/64	1.6	0.8	2.1
1860/64-1895/99	0.4	0.2	0.9
Sweden			
1861/65/1891/95	1.7	1.0	1.6
1896/1900-1926/30	1.0	0.4	0.8
Japan			
1878/82-1913/17	2.0	0.8	2.1
Mexico			
1940-60	5.1	2.2	--

[a]1829-59.

[b]Effective years--the eight-year period of war and reconstruction, 1941-48, is excluded.

[c]Data not available.

Source: Charles K. Wilber, "The Role of Agriculture in Soviet Economic Development," Land Economics, February 1969, pp. 91-92.

TABLE 2.31

Average Annual Growth Rates of Agricultural Output per Unit of Land (in percent)

Countries	Total
Soviet Union	
1928-1964/65	0.5
1928-40 and 1948-1964/65*	0.7
Grains, potatoes, vegetables, and industrial crops	
1929-59	0.9
1928-40 and 1948-59*	1.2
United States	
1910-40	0.7
1928-59	1.1
Japan	
1878/82-1913/17	1.2

*Effective years--the eight-year period of war and reconstruction, 1941-48, is excluded.

Source: Charles K. Wilber, "The Role of Agriculture in Soviet Economic Development," Land Economics, February 1969, p. 93.

TABLE 2.32

Agricultural Progress: Index Numbers for 1970 (1952 = 100)

	Agricultural Production		Food Production	
	Total Volume	Per Capita	Total Volume	Per Capita
World	165	115	170	120
Africa	160	105	160	105
North America	130	100	140	105
South America	165	105	170	105
Asia*	165	115	170	115
Europe	165	145	170	145
Oceania	185	125	190	130
Soviet Union	225	170	225	175

*The index does not cover China (mainland), Mongolia, North Korea, and North Vietnam due to lack of information.

Note: Certain extrapolations were necessary to bring the indexes in this table up to 1970 and the data should be considered preliminary and approximate. Numbers are rounded to the nearest 5 points.

Source: UN Statistical Yearbook 1969 (New York: United Nations, 1970), pp. 22-23.

That productivity of agricultural labor is higher in the United States than in the USSR is likewise explainable to a great extent by the advantages the former has over the latter: better climate, more fertile soils, more agricultural equipment, more fertilizer per acre, a more developed transportation system, more experience in the application of modern technology to agriculture, and so on. As to the much discussed factor of personal incentive, there is no doubt that the extent of motivation does make a difference and that at this stage of development, material incentives still provide the primary stimulus. But does this necessarily mean that the essential incentives must come from the private ownership of the means of production, even if it did play a major role in the early days of U.S. economic development, when privately owned businesses were small and owner-operated? After all, most of the rather efficient nonagricultural U.S. workers today are in the employ of impersonal corporations and neither own nor aspire to own the factories, railroad yards, and supermarkets, nor the schools and laboratories in which they work. In U.S. agriculture, 95 percent of all farms are still privately owned and operated; but these account for only 60 percent of sales of agricultural products in the United States, and for a mere 14.5 percent of the number and 8.4 percent of the sales of farms with annual sales of $100,000 or more. Overall, 80 percent of all farm marketings in the United States comes from 25 percent of the farms engaged in the production of the major food and fiber crops such as grains and cotton--those very products of commercial agriculture which have been the focus of U.S. agricultural policy over the past four decades and which are now the major crops in world trade.[66]

In the Soviet Union, no efforts have been spared in recent years to motivate collective and state farmers to put forth their best efforts. While nonmaterial rewards (medals, titles, honors) have not been neglected, the emphasis has been on material incentives. Special bonuses for workers on state farms which fulfill and overfulfill the plan, the extension of minimum income guarantees and of social security benefits to collective farmers, the tying of collective farmers' incomes to "workday units" based on type and quantity of work performed, the constant increase in procurement prices[67] combined with retail prices that have remained unchanged for the past 15 years, and ever increasing investments in agriculture are all clear evidence that the Soviet leadership has gone all out to subsidize agriculture and to stimulate agricultural performance. As a result of these efforts--and probably primarily as a result of the rapidly increasing investments--labor productivity in agriculture has greatly increased in the Soviet Union so that today, with about half as many people on farms, the Soviet Union's agricultural output is two and a half to three times what it was before the Revolution. The United States, to be sure, produces a somewhat larger agricultural output than the Soviet Union with one-eighth as many workers,[68] which puts it far ahead of the USSR in terms of labor productivity in agriculture. Although the agricultural labor force in the USSR keeps declining consistently, and although agricultural output has been rising much more rapidly in the Soviet Union than in the United States (from 1950-56 to 1966-71, for instance, by 84 percent in the USSR as compared with 32 percent in the United States),[69] the gap can be narrowed but gradually and because of previously discussed natural differences (climate and soil, primarily) will probably never be closed completely.

Ever since Stalin's collectivization drive, Soviet farmers have had the right to a private household plot, nowadays about half a hectare in size, on which they can raise food, including livestock, for their own consumption and for sale on what amounts to the only free market in the USSR.* On these private plots, which constitute 3 to 3.5 percent of the Soviet Union's total agricultural land area, is produced upward of 25 percent of its total agricultural output. Labor productivity, measured in terms of hours of labor time, is low on these private plots; but output per acre of land is obviously extremely high. Until earnings of Soviet farmers for their work on collective and state farms are much higher than they are today, such subsidiary plots will be welcome since they provide fresh food (vegetables, milk, eggs, and so on) for the farmers' table, as well as supplementary cash income. That small family-operated farmsteads are ill-suited for modern, mechanized production of such staples as grain, cotton, or sugar goes without saying.

In terms of technology employed in agriculture, the USSR also lags far behind the United States. Fertilizer is used more intensively and the ratio of mechanized farm equipment to cultivated area is much higher in the latter than in the former, all of which plays a major role in making agricultural land and labor more productive in the United States. But with Soviet investments in agriculture currently at a rate several times that in the United States, it is in the area of capital deficiency that the gap is likely to be narrowing most rapidly in years ahead. Overall capital-output ratios in agriculture (ratios of total capital stock in agriculture to total agricultural output) are much higher in the United States; but returns on additional investment are running low in the Soviet Union which, added to the large labor inputs, makes Soviet agriculture relatively high-cost.

---

*City workers can also have such a private plot if they really want to spend their free time cultivating it. A relatively small percentage of city workers take advantage of this opportunity.

In part, the low returns are simply due to the magnitude of such investments in recent years which necessarily had to result in diminishing returns. In part, however, the fault lies surely with investment decisions that are ill-conceived or inadequately coordinated--and the Soviets are the first to admit it and to strive for improvements. However this may be, many recent investment decisions seem to have been economically sound; and it is in the area of well-thought-out investments that the greatest hopes for rapid and continuous progress in Soviet agriculture lie. With Soviet farming operations highly labor-intensive in an economy encumbered by the persistence of an overall labor shortage, it would seem absolutely essential for more efficient operation that the inventories of mechanized agricultural equipment be increased. And, indeed, these inventories have been rising substantially in recent years and are scheduled for considerable increases again during the present 1976-80 five-year plan period. Land reclamation entails very high initial capital investments; but it often appears to have worthwhile long-run effects since by now drained and irrigated lands, which occupy only 6.5 percent of the USSR's plowlands, account for 24 percent of the country's agricultural produce.[70] Especially in light of greatly diminished opportunities for extensive growth elsewhere, it would therefore seem to have been rational that investment funds should have been allocated in recent years for the reclamation of millions of acres of waterlogged and of arid lands, and that the current five-year plan calls for the further millions of acres to be reclaimed by 1980. High priority is being placed on increasing the production of fertilizer, which is likely to pay off because at this point, Soviet studies show, each additional 100 kilograms (standard units) of mineral fertilizer applied will on the average provide additional yields of 100 to 200 kilograms for grain and 600 to 1,200 kilograms for sugar beets, potatoes, and vegetables, and because, the Soviets assert, lands will yield 2.5 to 3 rubles worth of additional farm product for each ruble spent on producing, transporting, and applying fertilizer to the fields.[71]

Surely, there are many shortcomings in Soviet agricultural investment which Western observers are quick to note and criticize. To increase the number of tractors doesn't do much good, they point out, as long as there are not enough spare parts to keep them in repair; total production of fertilizer has already passed that of the United States, but without sufficient equipment to transport and spread it, much is lost before it reaches the fields; due to inadequate transportation facilities--to roads impassible much of the year because of snowdrifts and mud and to motor vehicles unsuitable to cope with such conditions--losses of agricultural products in transit tend to be heavy and much perishable produce doesn't reach the cities at all; and there are not enough storage facilities so that, in good crop years especially, much grain has to be stored in primitive sheds on the farms and occasionally in open air with resulting extensive spoilage. This became particularly evident in 1973 when available storage capacity of 98.5 million tons on collective and state farms[72] proved totally inadequate for a bumper grain crop of more than 222 million tons.*

While these and other criticisms are surely well taken, one must not assume that the Soviets are ignorantly unaware of such shortcomings. In recent years especially, shortcomings have been discussed openly; and many corrective measures, if not on their way yet, are at least in the planning stage. After all, it was, for instance, Brezhnev who pointed to fertilizer losses in transit and called for improvement of the situation; citing another example, the Soviets have been fully aware of the need for additional storage facilities (caused, by the way, largely by increasing grain outputs), and the current five-year plan calls for the commissioning by 1980 of 30 million tons of new elevator capacity, twice the amount of the last five-year plan. Part of the trouble is of course that Soviet investments in agriculture are already running very high and that it is simply not possible to allocate funds for all desirable investments and to remedy all shortcomings in one, or for that matter in five or ten, years. Undoubtedly, shifting some investment priorities might well be economically advisable. It would, however, take a much more detailed study before specific questions could be answered with any degree of certainty--questions such as whether additional storage facilities or additional reclaimed land areas would yield higher economic returns, whether at this point in time additional investments in Kazakhstan or in the Ukraine would be more productive, or whether an increase in procurement prices or experimentation with the zvenos (small, independent, farm production teams) would be more likely to enhance farmers' motivation. The Soviets have in recent years become economically much more aware and there is every indication that they are giving due consideration to these and many other such questions.

And what about the future of Soviet agriculture? It is probably safe to say that the problems that have long beset Soviet agriculture will not be solved over night, but that the USSR's agricultural sector will still continue to play a major role in the Soviet Union's economic

---

*This does not mean that collective and state farms needed storage facilities for the entire 222 million tons. Some of the grain could be processed before all was harvested; some was quickly turned over to the state; and some was distributed to the collective farmers.

development and in the rise of the living standards of the Soviet people; that in time it will gradually become ever more efficient and productive; and that the gap between Soviet and U.S. agriculture will be narrowing in all respects. The Soviet Union might soon be able to surpass the United States in total agricultural output; but because of natural disadvantages, it might never be able to do so in agricultural labor productivity or in output per unit of land. For this, the Russian Revolution can hardly be blamed because, in the worlds of the well-known Western Sovietologist Alec Nove, "It would be absurd to imagine that Russian agriculture would approach American levels of productivity if Czar Nicholas IV were on the throne."[73]

NOTES

1. For these and similar statements, see, for instance, Ayn Rand, Capitalism, the Unknown Ideal (New York: Signet Books, 1967), p. 29; Roy D. Laird and Betty A. Laird, Soviet Communism and Agrarian Revolution (Baltimore: Penguin Books, 1970), p. 10 and passim; Alec Nove, "Will Russia Ever Feed Herself," New York Times Magazine, February 1, 1976, p. 9.

2. Quoted from Joseph Stalin's Works, vol. 12, p. 40, in Alec Nove, An Economic History of the USSR (London: Allen Lane, The Penguin Press, 1969), p. 188.

3. Alec Nove, "Was Stalin Really Necessary?" in Harry G. Shaffer, The Soviet System in Theory and Practice (New York: Appleton-Century-Crofts, 1965), p. 66.

4. USSR Agriculture (Moscow, 1975), p. 22. This official Soviet source estimates that in 1921 overall agricultural output was a mere 60 percent of the prewar level, crop production a mere 55 percent, and cattle herds had been reduced by 40 percent.

5. For detailed figures by regions see G. Konyukhov, KPSS v borbe s khlebnymi zatrudneniyami v strane (Moscow, 1960), pp. 64-65.

6. Programme of the Communist Party of the Soviet Union (Moscow, 1961), p. 64.

7. From an article by V. Matskevich [originally published in Nauki i Zhizn, no. 7 (1965)] in The Soviet Economy, 2nd ed., ed. Harry G. Shaffer (New York: Appleton-Century-Crofts, 1969), p. 156.

8. Current Economic Indicators for the U.S.S.R., Joint Economic Committee, U.S. Congress (Washington, D.C.: Government Printing Office, 1965), p. 71.

9. National Economy of the USSR (Moscow, Foreign Languages Publishing House, 1957), p. 41 gives the index for industrial output for 1940 as 583, with 1928 = 100.

10. With 1928 = 100, Norman M. Kaplan and Richard Moorsteen give the index for 1940 as 263 in "An Index of Soviet Industrial Output," American Economic Review, June 1960, p. 296; Richard M. Moorsteen and Raymond P. Powell give the index for 1940 as 291-318 in The Soviet Capital Stock, 1928-1962 (Homewood, Ill.: Richard D. Irwin, 1966), pp. 622-24. (The index number 291 is in 1955 prices, 318 in 1937 prices.) G. Warren Nutter gives the index as 312 in Growth of Industrial Production in the Soviet Union (Princeton, N.J.: Princeton University Press), p. 158; Naum Jasny as 350 in The Soviet Economy During the Plan Era (Palo Alto, Calif.: Stanford University Press, 1951), p. 17; Donald Hodgman gives the index as 430 in Soviet Industrial Production, 1928-51 (Cambridge, Mass.: Harvard University Press, 1954), p. 89; and Francis Seaton gives the index as 462 in "The Tempo of Soviet Industrial Expansion," Manchester Statistical Society, January 1957, p. 30.

11. Naum Jasny, The Socialized Agriculture of the USSR: Plans and Performance, Grain Economic Series No. 5 (Stanford, Calif.: Stanford University Press, 1949), p. 33. In this paragraph, the author leaned heavily on Charles K. Wilber, "The Role of Agriculture in Soviet Economic Development," Land Economics, February 1969, pp. 88-90.

12. With 1928 = 100, official Soviet sources give the index number for 1940 as 113.7. Narodnoe khoziaistvo SSSR v 1963 godu (Moscow, 1964), p. 227; Johnson and Kahan give the index as 120.3 in 1926-27 prices and 118.0 in 1958 prices in D. Gale Johnson and Arcadius Kahan, "Soviet Agriculture: Structure and Growth," Joint Economic Committee, U.S. Congress, Comparisons of the United States and Soviet Economies (Washington, D.C.: Government Printing Office, 1960), Pt. I, p. 204; Moorsteen and Powell give the index in 1937 prices as 120.3, op. cit., p. 622.

13. Kaplan and Moorsteen, op. cit., p. 296.

14. Johnson and Kahan, op. cit., p. 210.

15. July 1941 statement by Goering, quoted in Lazar Volin, A Century of Russian Agriculture (Cambridge, Mass.: Harvard University Press, 1970), p. 296.

16. Narodnoe khoziaistvo SSSR v 1958 godu (Moscow, 1959), p. 314.

17. Istoryia velikoi otechestvennoi voiny, 1941-45, vol. VI (Moscow: Institut Marksizma-Leninizma, 1960), p. 45; Moorsteen and Powell, op. cit., p. 624.

18. Kaplan and Moorsteen, op. cit., give the index of food production, 1927-28 = 100, as 163.7 for 1940 and 74.0 for 1945.

19. Promyshlennost' SSSR (Moscow, 1964), p. 142.

20. Nancy Nimitz, "Soviet Agricultural Prices and Costs," Joint Economic Committee, U.S. Congress, Comparisons of the United States and Soviet Economies (Washington, D.C.: Government

Printing Office, 1960), p. 254 gives the price index, with 1937 weights and 1928 = 100, as 461 for 1937, as 1,291 for 1948, and as 1,000 for 1952.

21. Narodnoe khoziaistvo SSSR v 1959 godu (Moscow, 1960), pp. 542, 546.

22. With 1940 = 100, G. Sorokin gives the 1950 index number as 99.5 in Planning in the USSR (Moscow, 1967), p. 177; and Norman Kaplan gives the index as 100 in The Record of Soviet Economic Growth, 1928-1965 (Santa Monica, Calif.: Rand Corp.), p. 5.

23. Sorokin, op. cit., p. 177; Narodnoe khoziaistvo v 1958 godu, p. 135.

24. Moorsteen and Powell, op. cit., p. 624; Kaplan and Moorsteen, op. cit., p. 296.

25. Programme of the Communist Party, pp. 62-63.

26. Volin, op. cit., Chapter 14, "Khrushchev: The Agrarian Leader and Reformer," p. 327.

27. See, for example, Douglas B. Diamond, "Trends in Output, Inputs, and Factor Productivity in Soviet Agriculture," Joint Economic Committee, U.S. Congress, New Directions in the Soviet Economy (Washington, D.C.: Government Printing Office, 1966), Pt. II B, p. 367.

28. USSR 75 (Moscow: Novosti Press Agency Publishing House, 1974), p. 200. See also, for instance, Narodnoe khoziaistvo SSSR v 1973 godu (Moscow, 1974), p. 392, which gives the grain output of Kazakhstan, where over half of the virgin lands are located, as 29 million metric tons for 1972 and 27.6 million for 1973, as compared with an average of 15.6 million for the first half of the 1960s.

29. Douglas B. Diamond and Constance B. Krueger, "Recent Developments in Output and Productivity in Soviet Agriculture," Joint Economic Committee, U.S. Congress, Soviet Economic Prospects for the Seventies (Washington, D.C.: Government Printing Office, 1973), p. 317.

30. See, for example, this author's "What Price Economic Reforms? Ills and Remedies," Problems of Communism, May-June 1963, pp. 19-27; idem, "Soviet Economic Growth and the Growth of Soviet Economic Rationality," Antioch Review, Summer 1966, pp. 187-203; idem, "East Europe: Varieties of Management," East Europe, November 1970, pp. 20-30, and January 1971, pp. 35-40; idem, "Planification et croissance économique: l'Union Soviétique et l'Europe de l'Est," Revue de l'Est, October 1971, pp. 75-122. As discussed in these articles, by the end of the 1950s, the system of detailed economic planning from the center had outlived its usefulness. It was ill-suited to cope with the problems of a mature socialist economy with its buyers' market, its need for intensive, rather than extensive, growth policies, and the astronomically increasing number of economic decisions it entailed.

31. Keith Bush, "Soviet Economic Growth: Past, Present, and Projected," Radio Liberty Research Supplement, September 12, 1975, p. 2. Annual average growth rates of U.S. GNP were given as 3.2 percent for 1951-60 and as 4.8 percent for 1961-65.

32. Twenty-fourth Congress of the Communist Party of the Soviet Union, Directives for the Five-Year Economic Development Plan of the USSR for 1971-1975 (Moscow, 1971), pp. 65-71; ibid., p. 30; Pravda, December 14, 1975.

33. Diamond and Krueger, op. cit., p. 316.

34. U.S. Central Intelligence Agency, The Soviet Economy in 1973: Performance, Plans, and Implications (Washington, D.C.: CIA, July 1974), p. 3.

35. Diamond and Krueger, op. cit., p. 32.

36. Pravda, December 14, 1975.

37. "Model Rules of the Collective Farm," USSR: Third Congress of Collective Farmers (Moscow, 1969), pp. 57, 61.

38. Keith Bush, "Agricultural Reforms Since Khrushchev," Joint Economic Committee, U.S. Congress, New Directions in the Soviet Economy (Washington, D.C.: Government Printing Office, 1966), Pt. II B, p. 456.

39. N. A. Borkhunov, "Nekotorye soobrazheniia o raspredelenii dokhoda v kolkhozakh," Finansy SSSR, no. 9 (September 1974), p. 38.

40. Pravda, December 3, 1975.

41. U.S. Central Intelligence Agency, Handbook of Economic Statistics 1975 (Washington, D.C.: CIA, 1975), p. 41 gives the 1965 population as 230.9 million and the 1974 population as 252.1 million.

42. Ibid., p. 40.

43. See David W. Carey, "Soviet Agriculture: Recent Performance and Future Plans," Joint Economic Committee, U.S. Congress, Soviet Economy in a New Perspective (Washington, D.C.: Government Printing Office, 1976), p. 582.

44. The Development and Allocation of Scarce World Resources, compiled by the Congressional Research Service, Library of Congress (Washington, D.C., May 1975), pp. 209-10.

45. Roy D. Laird, "Soviet Agriculture in 1973 and Beyond in Light of United States Performance," The Russian Review, October 1974, p. 383; The Soviet Economy in 1973, p. 5.

46. Comment by David M. Schoonover, Leader, Centrally Planned Countries Program Area, Foreign Demand and Competition Division, Economic Research Service, U.S. Department of Agriculture. Mr. Schoonover was a discussant of an earlier version of this chapter when it was delivered at the 1975 meetings of the Eastern Economics Association.

47. U.S. Central Intelligence Agency, The Soviet Grain Balance, 1960-73 (Washington, D.C.: CIA, September 1975), p. 18.

48. "Soviet Economy Today and Tomorrow," Socialism, Theory and Practice, August 1974, p. 133.

49. SSSR v tsifrakh v 1974 godu (Moscow, 1975), p. 120; Agricultural Statistics 1973 (Washington, D.C.: Government Printing Office, 1973), p. 433.

50. U.S. Central Intelligence Agency, USSR Agriculture Atlas (Washington, D.C.: CIA, 1974), p. 21.

51. Ibid., p. 21.

52. Handbook of Economic Statistics, 1975, pp. 20, 24.

53. F. Douglas Whitehouse and Joseph F. Havelka, "Comparisons of Farm Output in the US and USSR, 1950-1971," Joint Economic Committee, U.S. Congress, Soviet Economic Prospects for the Seventies (Washington, D.C.: Government Printing Office, 1973), p. 354.

54. Diamond and Krueger, op. cit., p. 334.

55. Izvestia, December 12, 1976.

56. Pravda, March 16, 1974.

57. Matskevich, op. cit., p. 167.

58. USSR Agriculture, op. cit., p. 96.

59. Nove, An Economic History of the USSR, op. cit., p. 24.

60. Wilber, op. cit., pp. 88-96.

61. USSR Agriculture, op. cit., p. 58.

62. Diamond and Krueger, op. cit., p. 335.

63. Laird and Laird, op. cit., p. 123.

64. Nove, "Will Russia Ever Feed Itself?" op. cit., p. 48. Computing the figure from other sources gives similar results. The 1971 Soviet wheat output was given as 98.76 million metric tons in Handbook of Economic Statistics 1975, p. 124; and the area sown to wheat as 158.1 million acres in Whitehouse and Havelka, op. cit., p. 346 (which is 63.98 million hectares). Dividing through gives a yield of 1.543 tons, or 15.43 centners per hectare. The wheat yield in the United States was 20.9 centners per hectare in 1970, 22.8 in 1971, and 22.0 in 1972 in U.S. Department of Agriculture, Agricultural Statistics 1973 (Washington, D.C.: USDA, 1973), p. 5.

65. Figures for the four East European countries from Nove, "Will Russia Ever Feed Itself?" op. cit., p. 48.

66. Committee for Economic Development (CED), A New U.S. Farm Policy for Changing World Food Needs (New York, October 1974), p. 20; U.S. Department of Agriculture, Family-Size Farms in U.S. Agriculture (Washington, D.C.: Economic Research Service, USDA, February 1972), pp. vi, 3.

67. By 1972-73, procurement prices were more than twice as high as in 1960 and several times as high as during the Stalin era. Borkhunov, op. cit., p. 38.

68. Whitehouse and Havelka, op. cit., p. 345, give the Soviet agricultural output in 1971 as 80 percent of that of the United States; U.S. Central Intelligence Agency, Handbook of Economic Statistics 1976 (Washington, D.C.: CIA, September 1976), p. 43, gives the 1975 agricultural labor force as 34.61 million for the USSR and 4.35 million for the United States.

69. Whitehouse and Havelka, op. cit., p. 345.

70. USSR 75, op. cit., p. 196.

71. USSR Agriculture Atlas, op. cit., p. 21.

72. Ekonomika sel'skogo khoziaistvo, no. 3 (1975), p. 47.

73. Nove, "Will Russia Ever Feed Itself?" op. cit., p. 48.

CHAPTER

3

HOW AGRICULTURE IS BECOMING AN ADVANCED SECTION OF SOCIALIST SOCIETY

Victor Perlo

The Soviet Union was the first country to introduce socialism, in agriculture as well as in industry. The problem in agriculture was immensely more complicated than that in industry because of the difference in class composition of industrial workers and peasants, and because of the tremendous lag in cultural and material conditions of the Russian countryside.

How successful has Soviet agriculture been, in terms of the overall goals of building an advanced socialist society in the Soviet Union, and in preparing for the prolonged task of constructing a not-yet-fully-defined communist society?

How successful has it been in serving as an inspiration and a rough model for the socialization of agriculture in other parts of the world?

I approach these questions from the criteria of Marxism, as developed in this century and referred to as Marxism-Leninism. And I consider them in a world of profound conflict between two classes, capitalists and workers, and two social systems, capitalist and socialist. Thus all questions must be studied comparatively. Success for socialist agriculture means doing better than capitalist agriculture.

Moreover, the approach has to be historical. The crudest aspect of capitalist propaganda about socialism is its use of selected static comparisons. This propaganda either omits, slurs over, or grossly distorts historical trends so as to convince the public of the inferiority of socialism through emphasis on the selected static comparisons.

With that general introduction, I pose the following specific questions:

How has Soviet agriculture contributed to the Marxist goal of eliminating the gap between city and country, between agriculture and industry, between worker and peasant?

How has Soviet agriculture contributed to the Marxist goal of overtaking capitalism in production per capita?

How has Soviet agriculture contributed to the fulfillment of the basic Marxist economic law of socialism: to increase steadily the material and cultural standards of the population, through constantly increasing production on the basis of the latest technique?

I will also deal with some special questions:

How did the rapid, intense, and often harsh collectivization drive of 1929-32 contribute to Soviet preparation for and salvation in World War II?

How did Soviet experience influence the socialization of agriculture in countries having socialist revolutions after World War II, and what has been the degree of success of socialist agriculture in these countries, in terms of the criteria stated above?

Wherever available, I use official Soviet statistics and official United States statistics. It is the custom of Central Intelligence Agency (CIA) researchers, and most academic students of the Soviet economy, to revise, and generally drastically revise, official Soviet statistics. The effect of these revisions is to make Soviet economic performance appear less successful, at any point in time and in trend.

Official statistics everywhere are apt to be distorted for propaganda purposes. I believe Soviet statistics, like U.S. statistics, are biased. However, the corrections made by CIA and similar statisticians in the United States often appear to be so extreme as to result in caricatures of reality, designed to satisfy crude antisocialist propaganda purposes.

I have the sixth edition of the famous textbook, Economics, by Paul Samuelson, published in the mid-1960s. Its cover pages have a chart purporting to show that economic growth has been faster in the United States than in the Soviet Union. To anybody with an even superficial knowledge of economic history since the foundation of the Soviet Republic, this is absurd, even fraudulent.

Scientific method would require correcting both U.S. and Soviet statistics for distortions, if one is going to seek that degree of precision. But the standard method is simply to make corrections, very great corrections, both absolutely and in trend, in Soviet statistics, while making no adjustments in U.S. official statistics.

For example, U.S. statisticians, in estimating Soviet per capita food supplies, often make adjustments for losses in transportation of Soviet crops. But they do not make adjustments for the very substantial undercounting of the U.S. population, which results in exaggerated measurements of per capita consumption.

On the whole, I have no reason to believe that Soviet statistics are any less accurate than U.S. official statistics. Moreover, I believe that the trends shown by Soviet statistics relating to agriculture are accurate, or at least not biased upward. Thus their use is consistent with a reasonable evaluation of the dynamics of Soviet-U.S. competition, and hence of socialist-capitalist competition in agriculture.

My thematic conclusion is that Soviet agriculture and the agriculture of other socialist countries which followed it have made the most rapid and dramatic progress ever made in agriculture under any social system. At the same time, I find that Soviet agriculture has faced many special difficulties, more intractable than those involved in the building of other parts of a socialist society. Thus Soviet agriculture has lagged behind the goals of government and Party leaders.

Through tremendous efforts by millions of people, and huge investments, the difficulties are being overcome, gradually and unevenly, but surely. Other socialist countries, starting later, benefiting from the initial assistance of the Soviet Union, from its experiences and mistakes, and with more favorable geographical and natural conditions, are progressing more rapidly and evenly in agriculture than the Soviet Union did in a corresponding stage of its development.

Under socialism, the farmer--historically the original capitalist owner of private property in the means of production and capitalist commodity producer--is adapting to cooperative, social forms of labor, and in so doing is gaining immeasurably in the breadth of his activity, his education, his standard of living, and all-around participation in the life of society.

SUCCESSFULLY TRAVERSING A DIFFICULT ROAD

Agriculture under czarism was backward technically and socially. Large parts of the country were dominated by big estates, farmed by poverty-stricken peasants, and powered by human muscles and insufficient draft animals. Only a generation away from formal serfdom by 1913, the peasants were little removed from that status in practice. The countryside lived on the narrow margin of starvation, with frequent famines in areas hit by crop failures.

Some areas, such as the Don and the Kuban, were farmed by Cossacks, essentially reserve soldiers given five hectare plots, and by landless and nearly landless peasants who had fled from Central Russia. It was in such areas that a class of more or less well-off peasants, the Cossacks, were dominant, but even these were decades behind the family-sized farmers of the midwestern United States in size of farm, amount of equipment, and political status.

On the eve of the 1917 Revolution, Russian agriculture had total power available of 23.9 million horsepower, of which 23.7 million consisted of horses and other animals, and only 0.2 million consisted of mechanical engines of all kinds.[1]

The land reform which accompanied and followed the 1917 Revolution distributed the land among more than 20 million peasant households and created a huge class of largely subsistence, independent peasants. Socially better off than before and able to feed themselves, the peasants by 1928 were little better equipped than in 1913, owing to the destruction of World War I and the prolonged Civil War, and the time taken to rehabilitate the ruined industry of the country and to replace the herds of livestock and horses.

Today, less than 50 years later, Soviet agriculture has more than halfway caught up to the United States in the supply of mechanical equipment, chemicals, and other industrial aids, and is being modernized in every way at a fast rate. Socially, it consists of large state farms and producers' cooperatives (collective farms), supplemented by farmers' personal plots. Increasingly it is being industrialized through large specialized farming, farm processing and servicing enterprises, run jointly by groups of state and collective farms.

To understand what was accomplished in the industrialization and modernization of the Soviet economy, consider that the Soviet government took over an overwhelmingly peasant country, much closer in living and technical standards to the Asian peasantry than to the West European and North American farmers. Even in 1928, on the eve of the five-year plans, the Soviet population was still about 80 percent rural, and nearly 80 percent of the economically active people worked in agriculture. The country was considerably less urbanized than Latin America.[2] But a half century later, Latin America, still almost wholly capitalist and largely under U.S. economic domination, remains a region of mass poverty and relative social and economic backwardness, while the Soviet Union is challenging the United States for social and economic world leadership.

In the United States, one has to go back to 1850 to find 80 percent of the population living in rural areas, and to 1830 to find 80 percent of the gainfully employed in agriculture.

By 1975 more than 60 percent of the Soviet population was urban. It took 47 years to get from 20 percent to 60 percent urban, while the United States was 110 years--until 1960--in effecting the same transition. Also by 1975, Soviet agricultural population was down to about 20 percent.[3] Again it took the United States 110 years to make the same transition, from 1830 to 1940.*

The victory of the Russian Revolution was based on an alliance of the working class and the bulk of the peasants. But the objectives varied. The working class--at least its politically active section--wanted socialism. The peasantry wanted land and petty commodity production. Both wanted an end of exploitation by capitalists and landowners. Up to a certain point they could help one another toward these different ends. But beyond a certain point, agriculture would have to shift to a socialist perspective if the country was to be successful in reaching its goals.

The basic class struggle part of the revolution in industry was completed rapidly, with the prompt socialization of all the main enterprises. But in agriculture the class struggle--aiming at the elimination of exploitation of hired labor by wealthy farmers--took about 15 years. And this prolonged struggle involved the majority of the population.

Soviet literature, including the great works of Sholokhov, describes vividly the class conflicts of the Civil War period and the later collectivization drive. The land reform which accompanied the early years of the revolution improved the situation of the poor peasants and of many "middle" peasants. This created a worker/peasant alliance, but simultaneously strengthened the private property consciousness of the peasantry.

Self-sufficient, near-subsistence farming became the rule; but capitalist elements gained ground, especially as better off farmers were able to buy land from the poorer farmers in exchange for needed animals, seed grain, and so on, and then to hire them as laborers at terribly low wages.

As a start toward introducing socialist forms, state farms were organized, and cooperatives of various types were encouraged. The cooperative movement spread, but mainly in selling and buying cooperatives, such as flourish even in capitalist countries. It took time for the peasants to see the advantages of the next step, producers' cooperatives.

---

*The figure of 20 percent of the labor force in agriculture in 1975 is based on the Soviet figure of 23 percent engaged in agricultural and forest economy in 1975, with a deduction for those engaged in forestry, and especially for the more than 3 million administratively attached to state and collective farms, but engaged in nonfarm work or in subsidiary industrial enterprises. The figure, therefore, is considerably lower than the 26.3 percent CIA estimate used by Shaffer.[4]

With the land divided into more than 20 million mainly small holdings, and many of these divided into several tiny strips, agricultural production increased rather slowly. The peasants, for the most part no longer exploited, lived much better than formerly, and so had less to sell on the market. The remaining 3 million rich peasants, or kulaks, controlled a large part of the marketable supply, and, with relatively little available for purchase in consumers goods, withheld much for both economic and political reasons.

Every step toward social cooperation in the countryside met with fierce opposition from the kulaks. This was no intellectual disagreement. It was violent opposition, featuring fires, slaughtering of livestock, and assassination of leaders and organizers. Throughout the 1920s, bands of dispossessed landlords and kulaks still on the land continued guerrilla warfare. There was plenty of incitation and practical assistance from capitalist intelligence services.

At the end of the 1920s, with industrial production and farm output ahead of the prewar level--although the marketable surplus of farm products was lower--the Communist leaders considered the time ripe for a major drive to industrialize the country and to mechanize and modernize agriculture. They considered this not only possible, but necessary for survival. They anticipated, correctly, that the contradictions of the surrounding capitalist states were leading to war, in which they might well be the prime target. Defense of the first socialist state would require a solid, modern, industrial base and a socially and economically reliable agriculture.

Collectivization of agriculture was necessary, then, for a series of interconnected reasons:

1. To permit application of modern farming methods and efficient use of machinery, which would be produced on a large scale under the industrialization program.

2. To provide an adequate surplus of farm products to feed the growing urban population, and for export in order to pay for imports of machinery.

3. To release surplus rural manpower for the new industrial and construction projects.

4. To bring closer ideologically the mass of peasants and the workers, and to put an end to class conflict in the countryside.

Stalin said:

> The way out is to turn the small and scattered peasant farms into large united farms based on the common cultivation of the soil, to introduce collective

> cultivation of the soil on the basis of new and higher technique. The way out is to unite the small and dwarf peasant farms gradually and surely, not by pressure but by example and persuasion, into large farms based on common, cooperative cultivation of the soil, with the use of agricultural machines and tractors and scientific methods of intensive agriculture. There is no other way out.[5]

By 1928, of 20 to 25 million peasant households, 10 million were in trading cooperatives, but only 800,000 in limited production cooperatives. The First Five-Year Plan provided for inducing 6 million peasant families, still a minority, to join producers cooperatives by 1933. These would be mainly poor peasants, with only 15 percent of the land. To increase marketable surplus, greater reliance was placed on formation of large state farms on virgin lands, and this part of the program was a prompt success, exceeding expectations in the initial period.

By supplying machinery and credits to collective farms of poor peasants it was hoped that they would increase their supply of grain, while the 10 percent of rich peasants would continue to account for one-third of the total marketed supply.

However, the kulaks saw this as a threat to their domination in the countryside, and greatly intensified hostile actions, sabotage, slaughtering of cattle, and murders of collective farm activists and Communist organizers. Forced to police action against the rich peasants, the authorities exerted excessive pressure on the middle peasants to join the collectives. These were formed, but not without the temporary alienation of some of their members, and not without big losses resulting from kulak depredations.

The slogan of eliminating the kulaks as a class was adopted. Property holdings deemed excessive were taken from them and turned over to the cooperatives. Some kulaks, who resisted most violently, were imprisoned or exiled to the East. Some stayed on as private farmers with reduced holdings and later joined the collectives. Others went to the towns and took jobs there.

Anti-Communist propaganda one-sidedly harps on repression of the kulaks. Meanwhile it keeps silent about the prolonged suffering of tens of millions under kulak exploitation, the terrible losses inflicted on the entire country by kulak sabotage, and the many who perished from hunger following the 1931 drought, the effects of which were compounded by kulak resistance. The fierceness of the struggle can be accounted for, not by the supposed ruthlessness of the Communists, but rather by the pressure of world capitalism, which still considered this first socialist revolution to be temporary and subject to destruction and which did everything to imbue in the remaining exploiting classes belief in the possibility of reversing the revolution.

Peasant membership in collective farms reached 14 million families by March 1930, but two-thirds of these dropped out after Stalin's famous "Dizzy with Success" speech in which he criticized forcible pressures on peasants to join the cooperatives. Thereafter, economic advantages offered by the cooperatives proved themselves and by the end of 1932 more than 60 percent of the peasants--14 million families--were permanently in the collective farms.

The beginning of the supply of tractors to the villages, especially to the state farms, was initially more than counterbalanced by the slaughtering of draft animals by kulaks and by middle peasants panicked by the kulaks during the first mass collectivization year.

With all the mistakes made, with all the attendant brutality, the campaign to convert agriculture to socialist forms was successful. By the late 1930s the collective farms were stabilized politically, their members were firm supporters of the Soviet system, and their production of crops stood at record levels, while livestock numbers had just about recovered the losses of the 1929-30 period.

Then came the war.

Since World War II the Soviet agricultural structure has gradually changed, with state farms increasing in importance and collective farms becoming more and more like state farms. The number of collective and state farms has been drastically reduced through amalgamations. In recent years, increasing attention has been paid to the organization of interfarm large-scale specialized agricultural enterprises, such as huge feedlots, and of industrial enterprises engaged in primary processing of farm products and other farm-related activities. This form is given high priority in the current Tenth Five-Year Plan. It may be that these enterprises will be the precursors of the reorganization of Soviet agriculture into a still smaller number of agrarian-industrial complexes, involving agricultural activity in a county-sized area, as in Bulgaria.

During the postwar period the gross production of Soviet agriculture increased as shown in Table 3.1. Production increased 2.5 times in 25 years, for a quite respectable average annual rate of increase of 3.8 percent. Since the number of agricultural workers, including those on private subsidiary plots, decreased 25 percent over this period,* gross productivity per worker increased 3.4

---

*Total employment in Soviet agriculture declined from 39.1 million in 1950 to 29.4 million in 1975, according to sources cited in Table 3.2.

TABLE 3.1

Postwar Agriculture in the Soviet Union

Years	Production in Billions of Rubles, Comparable Prices (average per year)
1946-50	35.7
1951-55	41.7
1956-60	59.2
1961-65	66.3
1966-70	80.5
1971-75	91.0

Note: Averages for five-year periods.
Source: SSSR v tsifrakh v 1975 godu (Moscow, 1976), p. 104.

TABLE 3.2

Agricultural Workers in the Soviet Union: Selected Years (in millions)

	Years			
	1940	1958	1965	1975
Total	41.9	40.2	32.2	29.4
Collective farm residents working on collective	26.4[b]	25.1	18.6	15.4
Working only on personal plot	6.6[b]	6.2	2.5	2.1[c]
Workers' family members on personal plot[a]	1.8	3.1	2.9	2.3[c]
Private farmers	4.7	-- [d]	--	--
State farm workers	2.4	5.8	8.2	9.6

[a]Includes family members of state farm workers mainly, but also family members of nonfarm workers who are occupied on personal farm gardening plots.

[b]Division between first and second column assumed proportional to 1958.

[c]Assumed equal to figures for 1974.

[d]Data not available.

Sources: Narodnoe khoziaistvo SSSR v 1964 godu (Moscow, 1975), pp. 390, 544, 545, 546; Narodnoe khoziaistvo SSSR 1974 (Moscow, 1975), pp. 448, 548, 549, 550; SSSR v tsifrakh 1975 godu (Moscow, 1976), pp. 134, 173.

times, for a compound annual rate of 5.0 percent. The increase in net agricultural productivity per worker, taking into account the increasing material inputs to agriculture, was considerably less.

The rate of increase in production has been faster than that sustained in capitalist countries over such a long period of time, including the United States. However, the rate of increase in labor productivity was almost exactly the same as that realized in the United States, and not exceptionally great, in this period of extremely rapid improvement in farm technology.[6]

The increase in farm production and productivity has been less than Soviet economic planners expected. It has been limited by special circumstances in the USSR, partly arising as the result of World War II and the subsequent military/political burdens of the country, and partly from shortcomings in agricultural economic policy and management.

The extraordinary potential of socialism to stimulate increased agricultural productivity is better illustrated by the experience of the German Democratic Republic. Although that country started at a considerably higher level than the USSR, net agricultural labor productivity increased at a compound rate of 5.6 percent per year between 1950 and 1974. (Details are provided under the section entitled "Model for Development.")

Harry D. Shaffer has noted that socialist countries with comparable climatic conditions achieved higher wheat yields than the United States in 1971-72. The same applies to those Union Republics of the Soviet Union with comparable climatic conditions. Thus, as compared with the 22 centners per hectare cited by Shaffer for the United States, the average yield in the Ukraine in the years 1971-74 was 29.7 centners, and in Moldavia 34.5 centners.[7]

EQUALIZING CONDITIONS BETWEEN TOWN AND COUNTRY

The abolition of social, cultural, and economic differences between farmers and workers, between town and country residents, has been a key goal of communist ideology. Friedrich Engels wrote in 1847:

> Industry which is carried on jointly and according to plan by the whole of society wholly presupposes people whose abilities have been developed all-around, who are capable of surveying the entire system of production. Consequently, the division of labor which makes one man a peasant, another a shoemaker,

> a third a factory worker, a fourth a stockjobber, already undermined by the machine system, will thus completely disappear.
>
> It follows from all this that the antithesis between town and country will likewise disappear. The carrying on of agriculture and industrial production by the same people, instead of by two different classes is, even for purely material reasons, an essential condition of communistic association. The scattering of the agricultural population throughout the country, alongside the crowding of the industrial population in the big towns, is a state adequate only to an undeveloped stage of agriculture and industry, an obstacle to all further development, which is making itself very perceptible even now.[8]

One marvels at the prevision of the founders of Marxist science who were able correctly to see so many stages ahead of their own time. Of course, the stage has not yet been reached where agriculture and industry are carried out wholly by people of the same social class, and with a similar quality of work, but, that stage is now already visible in the not-too-distant future of the Soviet economy.

Under capitalism, especially in this century, technical differences between agriculture and industry have been reduced in the industrialized countries, but serious economic and social differences remain. And in the developing countries, the hinterland, so to speak, of world capitalism, the contradictions between city and country are, if anything, more acute than a century ago. The "idiocy of rural life" of which Marx wrote is compounded by extreme poverty, a marginal and insecure existence, worsened by class exploitation, international price differentials against the rural producers, and deteriorating population/resource ratios.

Soviet policy, from the start, aimed to diminish and ultimately eliminate the differences in living standards, cultural levels, and quality of labor between town and country, between workers and farmers. Lenin wrote:

> Clearly, in order to abolish classes completely, it is not enough to overthrow the exploiters . . . it is necessary to abolish the distinction between town and country, as well as the distinction between manual and mental workers. This requires a very long period of time. In order to achieve this an enormous step forward must be taken in developing the productive forces; it is necessary to overcome

> the resistance (frequently passive, which is particularly stubborn and particularly difficult to overcome) of the numerous survivals of small-scale production; it is necessary to overcome the enormous forces of habit and conservatism which are connected with these survivals.[9]

And this was especially difficult in the Soviet Union. Lenin told a Party congress in 1921:

> There is no doubt that in a country where the overwhelming majority of the population consists of small agricultural producers, a socialist revolution can be carried out only through the implementation of a whole series of special transitional measures which would be superfluous in highly developed capitalist countries where wage-workers in industry and agriculture make up the vast majority. . . .
>
> . . . it will take generations to remould the small farmer and recast his mentality and habits. The only way to solve this problem . . . is through the material basis, technical equipment, the extensive use of tractors and other farm machinery and electrification on a vast scale. This would remake the small farmer fundamentally and with tremendous speed. If I say this will take generations, it does not mean centuries. But you know perfectly well that to obtain tractors and other machinery and to electrify this vast country is a matter that may take decades in any case. Such is the objective situation.[10]

It has taken four decades so far, in addition to time lost on account of wars and reconstruction from wars. Very substantial progress has been made. There is still a rather long way to go; but it is reasonable to project that before the end of the twentieth century the social and technical transformation called for by Lenin will be essentially complete, although, of course, further stages of progress will loom ahead.

Let's consider first economic conditions. Even in czarist Russia, workers, despite their extreme exploitation and poor living conditions, were far better off than the bulk of the peasantry. Moreover, with the overwhelming majority of the population peasants, to lift their living standard was a much bigger job than to lift that of the comparatively small working class which took power in the Russian Revolution.

Important progress has been made toward equalizing incomes between town and country. Part of the progress has been through the growth in importance of state farms, and the approach of farm workers' wages to those of city workers.

In 1940, wage and salary workers were only 6 percent of those engaged in Soviet agriculture. By 1958, they were 15 percent of the total, and by 1975, about one-third of the total. Their earnings have increased rapidly and since 1960 closed much of the gap below those of nonfarm workers.

TABLE 3.3

Average Wages and Salaries for Soviet State Farm Workers and Employees and All Workers and Employees (rubles per month)

Year	State Farm	All	Percent State Farm of All
1940	22.0	33.1	66.5
1960	53.9	80.1	67.3
1965	74.6	96.5	77.3
1970	100.9	122.0	82.7
1975	127.5	146.0	87.3

Sources: Narodnoe Khoziaistvo SSSR v 1964 godu (Moscow, 1975), p. 555; SSSR v tsifrakh 1975 godu (Moscow, 1976), p. 180.

In 1960, wages of state farm workers were 33 percent below the average of all workers; in 1965, they were 23 percent lower; and in 1975, only 13 percent lower. During these 15 years of stable retail prices, wages of farm workers have gone up 137 percent, faster than that of any other major industrial group. They are now equal to or above most trade and service group wages, but still considerably below the wages of industrial workers (162 rubles in 1975).

However, a Soviet economist estimates that nearly 20 percent of the total income of the average farm workers' family is from the family personal plot,[11] which would bring the real family income about up to that of a nonfarm family with the same number of workers. Of course, a considerable amount of extra labor has to be put in for this "moonlighting," either by working family members or dependents.

This near equalization of wages of farm and nonfarm wage earners is a unique accomplishment of socialism. It has gone further in the German Democratic Republic, where average monthly wages of farm and forest workers reached 859 marks in 1974, just a shade below the average for all workers of 867 marks, and for industrial workers of 865 marks. As in the Soviet Union, the wages of agricultural workers showed the fastest rate of increase.[12]

In capitalist countries farm laborers are the most exploited, oppressed section of the working class. In the United States, in July 1976, average hourly wages of farm workers were $2.53, as compared with $4.84 for private nonfarm production workers, or only 52 percent as much.[13] A more comprehensive set of statistics for 1975, compiled by the U.S. Department of Commerce, gives farm workers and supervisors $2.32 per hour, or 42 percent of the average of $5.57 per hour for all workers and employees.[14] (For annual income comparisons, see the end of this section before "Feeding the Population.")

These statistics for farm workers are derived from U.S. Department of Agriculture surveys, covering mainly large-scale employers, and perhaps excluding the bulk of the migratory workers, especially those without official entry documents ("illegal workers") who make up a large part of the U.S. farm labor force. These workers, often paid indirectly through labor contractors, receive miserly food and filthy lodging, and little or no cash after paying for the basic necessities supplied them.

The horrible conditions of migratory farm laborers in the United States--as in many other capitalist countries--are often exposed in press reports and legislative hearings, but have not been essentially changed over the years.

Farm workers in the Soviet Union are covered by the same 70 ruble minimum wage as all other workers and employees engaged in material production, and receive the same social security and vacation benefits. In the United States, only 43 percent of farm workers are covered by minimum-wage regulations. They are excluded from social security coverage, including old-age pensions. Nor do they receive unemployment insurance during the long periods of unemployment to which most of them are subjected. There is no such problem for Soviet farm workers.

Incomes of collective farmers have been drawing closer to those of workers, but remain significantly lower. Table 3.4 shows average monthly income of collective farmers, in money and kind, from their work on the collective farm, but not including their income from personal subsidiary plots.

Between 1965 and 1976, the average income of collective farmers increased 91 percent, while average wages and salaries increased

55 percent, and average collective farm income has increased year after year, even in years of declining farm output, such as 1975. Through the current system of monthly quasi-wage payments to collective farmers, they are in effect guaranteed a minimum monthly income, regardless of the effect of the weather or other conditions on ultimate results.

TABLE 3.4

Average Monthly Income in the Soviet Union
(in rubles)

Year	Collective Farmers	Workers and Employees	Percent Farmers of Workers
1965	51.5	96.5	53
1970	75.0	122.0	61
1975	93.5	146.0	64
1976*	98.2	149.5	66
1980*	116.0	170.0	68

*Planned.

Sources: N. K. Baibakov, in Ekonomicheskaya Gazeta, no. 50 (Moscow, 1976) (State Plan for 1976); N. K. Baibakov, in Pravda, October 28, 1976 (State Plan for 1976-80); SSSR v tsifrakh 1975 godu (Moscow, 1976), pp. 134, 135, 180.

The total real family income of collective farmers' families increased somewhat more slowly than income from collective farm work because of the relative decline in income from personal subsidiary plots. In 1974, collective farm income accounted for 44.2 percent of the family's real income; the personal plots added 26.6 percent; pensions, stipends, and other forms of social income, including education, added 19.9 percent; wages and salaries of family members 7.8 percent; and other sources 1.5 percent.[15] The relatively large income still received from the subsidiary plots was sufficient to bring the real income of a collective farm family up to that of a worker's family, provided the former had the same number of collective farm or other workers as the latter. In fact, the average collective farm household had only 1.15 collective farm workers, while the average worker's family may have had 1.75 to 2 workers.[16]

The lower income of collective farmers, as compared with state farm workers, is associated with their lower productivity and with less time spent on the collective farm. In 1974, the collective farmer spent an average of 19.9 days per month working in the cooperative, while the state farm worker spent an average of 23 days on the job. This accounted for about half the difference in their monthly earnings. Daily earnings of collective farmers were 4.50 rubles, 17 percent less than the 5.42 average of state farm workers.[17]

Soviet policy aims to reduce, and essentially eliminate, the difference between state and collective farms. A Soviet economist writes:

> The conversion of collective farms to a guaranteed wage geared to the basic wage scales of state farms and the development of material incentive funds on state farms upon their conversion to cost accounting have led to significant equalization of the sizes and the terms of payment of wages for collective farmers and state farm personnel. The state farm principle of the guaranteed wage has been introduced on collective farms; and with the conversion of state farms to full cost accounting, the wages of state farm personnel through collective incentive funds depend more and more on the income of the entire farm, which was previously the case on collective farms. The system of pensions and social security for collective farmers more and more approximates the conditions existing at government enterprises. The development of agrarian and agroindustrial integration is increasingly obliterating differences between collective farms and state farms.[18]

Taken as a statement of direction, of a change in process, this is accurate. But the change is far from complete. Soviet Premier Alexei Kosygin claimed:

> The life of the rural population has changed beyond recognition. Today electricity is used by 99 per cent of the collective farm households and homes of state farm workers and employees. Many tens of thousands of rural population centers now use gas. The majority of the rural population now has the possibility of using television sets. Almost every

> other working rural inhabitant has a higher or secondary (complege or incomplete) education. The incomes of the rural population are growing at a rapid rate. The conditions of life are in many ways drawing closer to those of the urban population.[19]

And yet, in many parts of the country, needed people are still leaving the farms for the cities, and this concerns most significantly the young people, the people with training, those with the greatest energy, who could contribute most to raising agricultural productivity. The primary reason here is the relative lack of supplies of goods and services in the countryside, the relative scarcity of cultural opportunities, and difficulties of access to urban centers. Where these problems are solved, agriculture is most successful. This is the case, for example, in specialized agrarian complexes near cities, where 25 percent of the workers actually commute from the cities.[20] Soviet people tell me that many Muscovites want jobs in such farms near that city, but that none are available.

But in large parts of the countryside dissatisfaction with conditions is still widespread, alcoholism is an evil and "vestiges of small-owner mentality prevail among some peasants to this day."[21] (In a later section I will discuss some of the difficulties and Soviet approaches to their solution.)

Equalization of conditions of cooperative farmers and workers has been largely completed in some other socialist countries, where agriculture was more advanced than in the Soviet Union at the end of World War II, where collectivization was less subject to political disruption, and where a more compact population geographically reduced the physical differences between town and country.

In Czechoslovakia, in 1970, for instance, per capita family income of blue-collar workers was 12,285 crowns, of white-collar workers 14,424 crowns, of cooperative farmers 15,033 crowns, and of pensioners 10,008 crowns.[22]

Equalization of economic conditions between farmers and workers has progressed much further in the Soviet Union and other socialist countries than in capitalist countries. In the United States, in 1974, the median earnings of male farmers and farm managers was $4,597, as compared with $9,064 for all male workers and $1,280 for male farm laborers.[23]

FEEDING THE POPULATION

How successful has Soviet agriculture, and the Soviet economy in general, been in feeding the growing population?

Any objective survey must conclude that the success was little short of sensational. In its 18-year respite between the Civil War and World War II, Soviet agriculture, the food-processing industry, and the transport and trade network ended the periodic famines and chronic mass malnutrition which had afflicted large sections of the population. They provided everybody with an adequate, if still poorly balanced, diet.

The destruction wrought by the German invaders and their allies in World War II brought per capita food supplies almost down to the 1913 level. But in the 30 years since, food consumption standards have been brought up to the general level of Western Europe. The diet is ample in calories, sufficient in proteins and other necessary ingredients, but still below the scientifically established optimum nutritional norms--moderately so in consumption of meat, substantially so in consumption of vegetables and fruit, as shown in Table 3.5.

This table shows a really remarkable performance. Consider the index of per capita food consumption, prepared by United States analysts, rising from 100 in 1950 to 159 in 1965 and 210 in 1975. Certainly no other large country, and perhaps no capitalist country of any significant size, has doubled per capita food consumption in such a period. Of course, this does not mean doubling food calories, but doubling food values, more than doubling consumption of meat and fish, increasing consumption of eggs nearly 3.5 times, fruits more than 3 times, and milk and vegetables 80 percent and 70 percent respectively.

Soviet food consumption is still below that of the United States in overall quality, but not decisively. In 1975, with calorie consumption somewhat larger in the USSR, per capita consumption of meat, poultry, and fish combined was 21 percent lower in the USSR, eggs 22 percent lower, milk 25 percent higher, sugar slightly higher, and fruits still far below the U.S. level (40 percent in weight, probably more than 50 percent below, with proper allowance for the food value of canned and frozen fruits and juices).

Note that the pace of improvement, in absolute terms, has been fully maintained well into the 1960s and 1970s. Thus, the total increase in meat and egg consumption in the 10 years from 1965-75 exceeded the total increase in the previous 15 years, and the per annum increase in milk and fish consumption was larger in the later period than in the earlier period. The U.S.-constructed overall index increased about 5 points per year in 1965-75, as compared with 4 points per year during the previous 15 years.

Yet, in what is still the most typical fashion of our Sovietologists, Gertrude E. Schroeder and Barbara S. Severin, who calculated these indexes, belittle the Soviet accomplishments and exaggerate the shortcomings with phrases such as a recent "slowdown in consumption," characterizing the gains as "typical of developing countries," the goals for the present five-year period as "modest," and estimating a decline of meat consumption, following the 1975 small grain crop, of "as much as one quarter."[24]

TABLE 3.5

Per Capita Consumption of Food Products in the Soviet Union (kilograms, except where otherwise noted)

	1913[a]	1950	1965	1975
Meat (including poultry)	29	26	41	58
Milk and milk products	154	172	251	315
Eggs (units)	48	60	124	215
Fish and fish products	6.7	7.0	12.6	16.8
Sugar	8.1	11.6	34.2	40.8
Vegetable oil	--[c]	2.7	7.1	7.9
Potatoes	114	241	142	120
Vegetables and legumes	40	51	72	87
Fruits and berries[b]	11	11	28	37
Grain products	200	172	156	142
Index of food consumption, (1950 = 100)	--[c]	100.0	159.1	210.0

[a]Consumption of food by working people in 1913 was considerably below the average. Source estimates that among working people in 1913 per capita consumption of meat was about 20 kg., and of milk about 120 kg.

[b]Excludes fruit and berries used for wine making.

[c]Data not available.

Sources: Narodnoe khoziaistvo SSSR v 1964 godu (Moscow, 1965), p. 603; SSSR v tsifrakh 1975 godu (Moscow, 1976), p. 202. Index of food consumption per capita from Gertrude E. Schroeder and Barbara S. Severin, "Soviet Consumption and Income Policies in Perspective," Joint Economic Committee, U.S. Congress, Soviet Economy in a New Perspective (Washington, D.C.: Government Printing Office, October 14, 1976), Table 1, p. 622.

How can one speak of "typical of developing countries" when the real course of development in the capitalist world during this quarter century has been a conspicuous widening gap between industrialized and developing countries, a continuation of chronic mass hunger in the latter, and a failure, taking the group as a whole, to make any decisive gains in per capita food consumption?

Through preserving some meat products, increased imports, and increased supply of fish, the Soviet trade network, apparently, has kept within moderate bounds the loss in consumption resulting from the need to rebuild livestock herds in 1976. During the first half of the year, sales of meat and meat products by state and cooperative stores declined 1 percent, while sales of fish increased 12 percent.[25] Full year declines in meat consumption may be somewhat wider, but nowhere near the 25 percent conjectured by Schroeder and Severin.

The claim has been published that Soviet per capita food consumption is still lower than U.S. per capita food consumption was prior to the Russian Revolution. And this may well be, if one considers average per capita consumption. By the early decades of this century, U.S. farmers were already applying machinery and other capital equipment to agriculture, benefiting from tremendous areas of arable land, much of it only recently under cultivation, and not yet exhausted or eroded.

Going back to 1850, when the United States had acquired approximately its present "conterminous" boundaries, the total population was only 29 million. In 1913, czarist Russia had a population of 159 million. But the two countries had about the same amount of arable land, and, as explained by Clark and Shaffer, climatic conditions were, as now, much less favorable in Russia. So czarist Russia had to feed six times as many people, with a technology little, if any, more advanced than that in use in the United States 60 years earlier. Add that the social system in the capitalist United States was more favorable to production than that in semifeudal czarist Russia. Thus, it is understandable that the United States, during most if not all of its history, has been able to feed its population better than the Soviet Union did for a large part of its relatively short history.

The fact of the matter is that capitalist economic contradictions and crises brought about a deterioration in U.S. food consumption for several decades after World War I, precisely when the socialist Soviet Union was laying the basis for its later rapid gains in consumption. In the United States, per capita meat consumption, slashed by inflation and depression, fell substantially below the pre-World I level for decades afterwards. In 1935, per capita consumption of meat was 28 percent below the 1908 level, which was regained only in 1962. Per capita consumption of fresh fruit continually trends downward, and is now about 40 percent below the average pre-World War I level. Per capita milk consumption is 30 percent below the pre-World War I level.[26]

Of course, there has been no decline in the food consumption of the well-to-do, and some of the reductions--for example, milk--have been partly for dietary reasons. But at the same time, the decline in per capita consumption of key foodstuffs suggests that the consumption standards of the poorer sections of the population have not improved, and may have significantly deteriorated with the expulsion of small farmers and sharecroppers from their subsistence patches on the land.

The continuation of malnutrition in the United States has been dramatized in recent years of inflation and economic crisis with widespread reports of poor people, especially elderly people, consuming dog food to obtain needed proteins, and congressional surveys revealing mass hunger among black people in some southern states. Undernourishment and poorly balanced feeding remains the fate of tens of millions because of racial discrimination and impoverishment of the low-income layers of the working class. Thus, in 1971-72, the median per capita calorie consumption of low-income black people was 19 percent below, and their median protein consumption 23 percent below, that of the entire population.[27] Roughly 40 million people receive some sort of official aid to food consumption--surplus food distributions, free school lunches and food stamps--under conditions which often imply inferior food quality and with the stamp of humiliating charity.

True, there are significant income and consumption inequalities in the Soviet Union. But the variations are less, they have no racial or national flavor, and partial equalization of food consumption is achieved through subsidized low-cost workplace and school meals, and subsidized basic food prices for all. The retail prices of meat and milk are subsidized by the Soviet government to the tune of the ruble equivalent of $25 billion per year, according to N. T. Glushkov, Chairman of the State Committee on Prices.[28] In addition, cash income subsidies to lower per capita income families are now granted by law in the USSR.

On this subject, what is really relevant is the following: In the United States, under capitalism, there are tens of millions whose food consumption is inferior to the average per capita food consumption here before World War I. In the Soviet Union, under socialism, virtually the entire population enjoys food consumption superior to the average per capita food consumption in czarist Russia before World War I.

This is not to decry the food and agriculture situation in the United States. U.S. agriculture, built up by the labor and skills of farmers and farm workers, the scientific work and inventiveness of selectionists, chemists, machinery designers, hydraulic engineers, and the great productivity of U.S. industry, is among the most advanced in the world. Still, shortcomings in the food supply are scandalous, considering the wealth of the country and the great productivity of U.S. agriculture. They are due to the capitalist system of production and distribution, and are not in the process of decisive improvement.

Shortcomings in food supply in the Soviet Union are due to natural difficulties, to the backward countryside inherited from the czarist regime, to man-made calamities imposed by anti-Communist invaders and protracted economic warfare, to policy mistakes of the Soviet leadership, and to bureaucratic obstacles to carrying out correct policies. But the tremendous progress made testifies to the fundamental merit of the socialization of agriculture, and to the overall improvement in specific policies and in their execution.

As with other aspects, one has to evaluate not just Soviet agriculture, but the efficacy of the socialist system, in which the USSR pioneered, in feeding the population of the countries where agriculture has been socialized. In the Eastern European countries this was done under more favorable conditions of climate and weather, and with less of a handicap imposed by World War II destruction, than in the USSR. Some of these countries, the Balkan countries and also Hungary and Poland, were poor, largely peasant countries before World War II, with a large part of the population suffering from malnutrition and undernourishment. Thirty years of socialism have brought dramatic gains in the food consumption of all of them, with some achieving the world's highest food consumption levels.

The German Democratic Republic, Czechoslovakia, and Hungary roughly approximate or approach the U.S. per capita consumption of meat, eggs, milk, and fruit. The Hungarian Minister of Agriculture, Pal Romany, recently claimed that in 1975 Hungarian per capita meat consumption was 308 pounds, as compared with 210 in the United States and 165 in the European Common Market countries.[29]

Bulgaria has become a major supplier of fruit, vegetables and animal food products throughout the socialist countries and also in Western European countries, while rapidly raising internal food consumption standards. Already Bulgarian food consumption is well ahead of the standards of Italy and other southern European capitalist countries, and far ahead of neighboring Turkey, from which Bulgaria won its freedom a century ago.

By 1980, the Bulgarians plan to reach excellent, balanced standards. Per capita meat consumption, which increased from 29 kilograms in 1960 to 41 kilograms in 1970 and 56.5 kilograms in 1975, is scheduled to go to 70 kilograms in 1980. Milk consumption, having increased from 92 kilograms in 1960 to 117 kilograms in 1970 and 142 kilograms in 1975, is scheduled to reach 220 kilograms in 1980. Egg consumption, after rising from 84 eggs in 1960 to 122 eggs in 1970 and 145 in 1975, is to reach 200 eggs in 1980. The already high levels of fruit and vegetable consumption are to increase further.[30]

The following figures compare per capita meat production of the industrialized capitalist countries with that of the socialist countries in 1975. Production came to 78 kilograms in the United States, 61 kilograms in the European community, 13 kilograms in Japan,[31] 60 kilograms in the USSR,[32] and 88 kilograms in the CMEA* countries of Eastern Europe,[33] this last figure being for 1974.

The figures for the capitalist countries have been taken from those compiled by the CIA. Those for the socialist countries come from their national statistical agencies. The CIA presents the same figure for the Soviet Union, but a much lower figure of 55 kilograms for the CMEA countries of Eastern Europe. This is clearly in error. The East European countries export substantial quantities of meat, but their per capita consumption was still a high 69 kilograms in 1974.[34] Inclusion of fish would considerably increase the per capita production figure of sources of protein for the USSR and among nonsocialist countries, especially for Japan.

According to this key measure of quality food production, the European socialist countries are now at least on a par with the industrialized capitalist countries (and would be, even according to the erroneous CIA figures). Of course, there are considerable variations among countries in each group, and consumption figures are affected by foreign trade balances. But the variations among the socialist countries are less marked than among the capitalist countries, and, in accordance with their principle of cooperation, the levels of meat consumption are evening out as they are increasing.

Within the European community, particular significance attaches to meat production in Italy, at only 37 kilograms per capita. Prior to World War II, most East European countries were less developed economically than Italy, while the latter was a substantial

---

*Council for Mutual Economic Assistance, the socialist country's counterpart to the European Economic Community (Common Market), although quite different in organization and functions.

food exporter. But the rapid industrialization in the socialist countries was accompanied by corresponding development in agriculture. In Italy, on the other hand, rapid industrialization after World War II, under conditions of capitalism, led to a stagnation of agriculture, and its deterioration relative to the needs of an expanding urban population. so that Italy has become a major food importer, and large sections of the population cannot afford an adequate diet.

United Nations statistics, arranged now according to social groups of countries, bring up to date comparative trends of per capita food and total agricultural production, as presented for the period 1952-70 in Table 2.32 and updated in Table 3.6.

TABLE 3.6

Index of per Capita Food and Agricultural Production
by Groups of Countries, 1974
(1961-65 = 100)

Groups of Countries	Food	Total Agricultural Production
World	108	107
Developed market economies	115	112
Developing market economies	99	99
Centrally planned economies	120	120
Asia	110	111
Europe and USSR	132	131

Source: U.N. Statistical Yearbook 1975 (New York: United Nations), p. 24.

There has been an increase in the industrialized capitalist countries, but the increase has been two to two-and-a-half times as fast in the industrialized socialist countries (Europe and the USSR). Per capita production has failed to increase, and has even declined slightly (significantly in Africa) in the developing capitalist countries, while it has increased definitely, if slowly, in the developing socialist countries (those in Asia). The largest developing socialist country, China, has not cooperated economically with the developed socialist countries in the last two decades, which has contributed to the relative slowness of its agricultural progress. But even so, its record compares quite favorably with that of the developing countries where capitalism still prevails.

SOCIALIST FARMING, ECONOMIC EFFICIENCY, AND THE WAR

Can a country develop a generally socialist economy for a long time while agriculture remains mainly private? Some light on this question is shed by reference to the Polish experience. This country approximated the Soviet model of building socialism in industry, social services, and international relations. But in agriculture, alone among the CMEA countries, farming remains overwhelmingly in private hands. Individual farming under socialist laws which forbid hiring labor and limit the size of individual holdings to what a family can care for with its own efforts is full of contradictions.

The farmer has his own land. He is able to feed his family. He shares in some of the social benefits of a socialist society. He is guaranteed stable prices for his products. He can get credit for the purchase of needed equipment. But inevitably, this approaches subsistence farming. In Poland, with its average farm of less than 12 acres, the marketable surplus remains small. And most of the tiny farmers lack the know-how and the space for really efficient farming, not to speak of specialization.

However, a Wall Street Journal correspondent distorts the picture when he writes: "The tiny size of farms in Poland, Eastern Europe's biggest agricultural producer, is at the heart of the country's difficulty in feeding itself, a difficulty that forces it to import Western produce, especially feed grains."[35]

The fact is that farm output in Poland has increased about as rapidly as in the other European socialist countries. The level and quality of the population's food consumption has improved remarkably. And generally, the country's exports and imports of farm and food products are about in balance. But its agriculture is relatively inefficient, with low labor productivity in comparison with some other socialist countries. Thus, the percentage of national income produced per person engaged in agriculture is 46 percent of the average for the entire economy in Poland, as compared with 78 percent in the USSR, and 92 percent in the German Democratic Republic (GDR).[36]

The stresses over food prices and shortages that occasionally break through the surface in Poland may be directly connected with the prevalence of privately controlled, dwarf-agriculture, and the power of the small farmers to withhold supplies. Also, private property ideology, generated by petty proprietor agriculture, may undermine the political solidity of the regime.

Yet, Polish agriculture benefits from the overall socialist control of the economy. This provides the private farmers with credits, access to equipment, technical assistance, and guaranteed stable prices. Thus, the private farmers are not subject to the failures which squeeze out small farmers in capitalist countries. This basically explains why farm output has increased, even while small holdings remain the basic form.

The Polish government promotes cooperative farming and sets up state farms. Also, intermediate forms of partial cooperation are encouraged. A substantial percentage of the farmers have become worker-farmers, working in industry part of the time and farming part of the time, a practice common in the United States. The government and the Polish United Workers Party, the leading political party, are determined to convert agriculture ultimately to socialist forms, but are extremely careful not to force the pace, so as not to alienate the farmers. They anticipate that socialization will accelerate as the older generation of small farmers retire.

The Soviet Union could not afford to wait so long to achieve socialist forms of agriculture. Without collectivization, the Soviet Union could not have released as much labor, and consequently could not have pursued industrialization and built up its infrastructure and defense establishment on the scale that it did. Problems similar to those faced by Poland after World War II were faced by Soviet agriculture in the 1920s--similar but worse, as there was no neighboring socialist country to help out with emergency shipments of grain when peasant production failed, or kulaks withheld deliveries, and no Warsaw Pact to restrain outside intervention in favor of counterrevolutionary forces.

Without previous socialization of agriculture, the Soviet Union could not have mustered the political and economic strength to defeat the powerful invaders in World War II. And one can go further: without having completed collectivization sufficiently in advance of the war to permit the consolidation of the collective farm system and the recovery of production from rich-farmer sabotage and arson, the USSR and its socialist industry and Communist-led governmental structure would probably have been destroyed.

Had the 3 million kulaks of 1928 remained in the countryside, Hitler, in 1941, might well have found the mass political support he counted on, might well have been able to convert the Ukraine into a fruitful granary and source of willing recruitment for his labor force. Instead he found almost unanimous opposition. Traitors to act as local repressive forces were few and weak, and generally untrustworthy for the Nazis. The collective farmers did all they could to prevent crops from falling into Nazi hands. They even delivered them across the front lines to besieged Leningrad. In more remote areas, they secretly retained the collective farm organizations. And hundreds of thousands joined in partisan or guerrilla warfare against the invaders, while the whole countryside helped feed and shelter the well-equipped Soviet forces airlifted to the rear of the invading armies.

The economic contribution of socialized agriculture was equally vital. The handicaps were seemingly overwhelming. By the late summer of 1942, the Axis powers had overrun regions accounting for about 60 percent of the current farm output of the USSR. In addition to the plundering and wrecking by the invaders, according to former Minister of Agriculture V. Matskevich: "large numbers of tractors, vehicles and horses were recruited for the front; there was a shortage of spare parts and fuel." Most of the able-bodied male farmers went to the front.

> The invaders destroyed and plundered 1,876 state farms, 2,890 machine and tractor stations, and 98,000 collective farms, wrecked or stole 137,000 tractors, 49,000 harvesting combines. . . . The invaders wrecked 285,000 collective farm buildings which maintained livestock, destroyed 1,262,500 acres of orchards and 382,500 acres of vineyards. Animal husbandry suffered particularly heavy losses: 7,000,000 horses, 17,000,000 head of cattle, 20,000,000 pigs and 27,000,000 sheep and goats were slaughtered or taken away to Germany.
>
> The losses suffered by the collective farms alone amounted to 181 billion rubles. . . .
>
> The war was an historical test to the collective farm system. During the trying war years Soviet farmers displayed a consciousness of social interests historically unprecedented.[37]

No invaded capitalist country ever experienced anywhere near such losses as were suffered by the Soviet Union and subsequently succeeded in recovering and fighting on to victory.

Socialist organization of industry, combined with the intense, unremitting labor of old people, women, and youth made it possible for the Soviet Union, with one-third the industrial power base of Germany and its occupied territories in Western and Central Europe, to match and then outproduce Germany in war materials by 1943.

And similar efforts and organization enabled badly crippled Soviet agriculture to provide spartan but adequate supplies for the Army and for the urban population. An official Soviet history of the war period stated:

> At many collective and state farms the number of able bodied workers had halved. There was a

> shortage of haulage power. The fields were overgrown with weeds. Yet thanks to the selfless efforts of the collective farmers and the assistance rendered by the entire country the difficulties were gradually surmounted. Most of the men had gone to the war and their places were taken by women, teen-agers and old men. . . . Women learned to operate harvester combines, tractors and other machinery. . . . A nationwide socialist emulation movement, along the lines of that in industry, was launched in agriculture.[38]

Another source added:

> Female labor in industry went up from 38 per cent in 1940 to 55 per cent in 1945. In agriculture the percentage rise was still greater. More than 250,000 rural women were in executive positions--collective farm chairmen or board members, team leaders and livestock farm managers. . . . Boys and girls of 16 and under comprised 17 per cent of the farmers in 1944.[39]

With all the effort, the grain crop shrank from 95.5 million tons in 1940 to 29.7 million tons in 1942, or by two-thirds. Moreover, although more land, recovered after the late 1942 battle victories, could be planted in 1943, the crop then was only 29.4 million tons owing to a drought. There was a 50 percent increase in the crop in 1944, and a modest further rise in 1945, but production remained well below the prewar level.[40]

The drop in production of 1942-43 would have led to mass starvation under conditions of private farming. As it was, however, the collective farms delivered the bulk of the grain grown to the State, so that the decline in grain deliveries was much smaller. In the four years of war, 1941-44, the state and collective farms delivered 68.8 million tons of grain to the State, three times as much as the 22.5 million tons delivered during the four years of World War I. G. Deborin writes:

> The food situation was difficult throughout the war. The stocks of grain, flour and cereals available at the beginning of the war (six million tons) were enough only for the first several months. The loss of agricultural areas was a bitter blow. Deliveries of grain, flour and cereals from the United States

> and Canada were negligible (adding about some 0.5 million tons in all the war years). Food supplies had to be strictly regulated. . . .
>
> Though generally coping with the food shortage, the city people suffered great hardships. Their diet shrank, and its quality deteriorated. In the cities, adult consumption declined from 3,370 calories per day to 2,555 calories, increasing only slightly to 2,810 calories in 1944. Consumption of fats, and especially of sugar, declined steeply, and that of proteins, too.[41]

That the peasants of Russia, as of many other countries, were patriotic and willing to defend their land against an invader even before the revolution was proven at the time of the Napoleonic invasion. But their performance during World War II justifies the conclusion that through the socialization of agriculture, a process in which the peasants themselves played an active part, they felt they had won a special stake in the revolutionary regime, had a special interest in defending it. And through the organizational possibilities of socialized farming, their contribution could be effective in a way that would be impossible in any system dominated by an exploiting, property-owning class, not to speak of the corrupt, bureaucratic, bumbling czarist regime. The English-language, one-volume history of World War II, recently published in Moscow, correctly concludes: "However, rigid rationing assured almost unintermittent supplies for the Army and the nation. The vitality of the collective and state farm system and the patriotic devotion of the peasants enabled the country to cope with the war-time difficulties."[42]

With this background, it is understandable that the 1975 drought, cutting grain production in the Soviet Union from 196 million tons in 1974 to 140 million in 1975, created problems that were serious enough, but mild in comparison with those that the Soviet system handled during World War II, and, for that matter, throughout the difficult decade thereafter. The import of a large quantity of grain from the United States was a convenience, avoiding the need to cut consumption standards moderately, but hardly, as some U.S. commentaries would have people believe, necessary to prevent hunger and social disorder in the Soviet Union.

MODEL FOR DEVELOPMENT

The Soviet system of state and collective farms has had an international, even a global, significance. It showed that peasants--

the traditional carriers of rugged individualism, of intense attachment to isolated private property in means of production, the original source of capitalist ideology, capitalist commodity production, capitalist exploitation of the weaker by the stronger, the propertyless by the propertied--could become integrated into the antithesis of capitalism, a socialist system of economy. They could merge their individual property holdings into cooperative property holdings with their neighbors. They could draw increasingly close to, and ultimately merge with, the workers and intellectuals of a society of publicly owned property, property of the entire people.

The first working class revolution in history, the Paris Commune, failed because the workers failed to form an alliance with the peasants who surrounded them and who constituted the majority of the population. The leaders of the Russian Revolution, influenced in part by the study of that experience, were firmly committed to the principle of the worker-peasant alliance in winning power and building socialism. They gave it special emphasis in their 80 percent peasant country, although they and other Communists recognized its global validity.

The forms of socialized agriculture developed in the Soviet Union slowly, and after a number of experiments that failed, such as early "communes" in which everything was pooled and all shared alike. Ultimately Soviet agriculture crystallized in two main forms, state farms and collective farms, or cooperative producers' artels.* The state farms employed initially very poor peasants and landless farm laborers who had the least property to be attached to, and for whom employment as workers by the State, under conditions superior to that of employment by rich farmers or landlords, represented an improvement in condition. Later, as the means and knowledge for scientific, highly productive agriculture became available, the state farms became advantageous places in which to work, with material conditions closely approximating those of city workers and more favorable than those on collective farms.

Collective farms included the masses of poor and middle-sized peasants, and those of the rich peasants and farmers who calculated, correctly, that it was futile and self-destructive to try to defeat socialism, and necessary for their own survival to give up privileges and join their poorer and often less efficient neighbors. Forms of the collective farms varied and developed dynamically with the general industrial advance of the country and the cultural advance of the collective farmers. Thus, in many early cooperatives, only the

---

*The specific form of cooperative farm which is standard in the Soviet Union.

land was worked cooperatively, while each peasant retained his own livestock. In some collectives, distribution of the proceeds was partly according to work, and partly according to the contributions of the different peasants in land, livestock, and equipment. The shift to complete pooling of collective property, and distribution strictly according to work came later in these cases.

Furthermore, in the advanced forms of artels which became the universal norm, each member retained or was given a supplementary plot, and was enabled to have personal livestock up to specified limits. The personal plots were and remain a normal component of state farms also.

These personal plots were a logical and necessary component of socialist agriculture in the Soviet Union. They were necessary ideologically, to provide a bridge between the traditional private-property ideology of the peasants and the ideology of collective ownership, between "mine" and "ours." They were necessary economically because at the existing level of mechanization and agricultural know-how, and of transportation and processing facilities, it was not possible to carry out the highly specialized agriculture that could make each state or cooperative farm a 100 percent commodity-producing enterprise, whose members would purchase their individual food on the national market.

Hence, the private plots provided the bulk of the food of the peasant families, and a surplus which they sold on the open market, either as individuals or through the agency of collective farm members who would take the produce of many individuals, as well as the collective farm's surpluses, to public markets in the towns and cities. A large part of the labor of the collective farmers was devoted to these personal plots. Part of the farmers' payment for work on the collective farm was in the form of grain for feed, seed for planting, and food for family consumption, of types that they did not produce on their own plots.

On the eve of World War II, the private plots provided more of the average collective farm family's incomein money and kind than their work on the collective enterprises. Since then, their importance had dwindled, relatively, but remains significant and is expected to play a substantial role for some time to come.

The Soviet state and collective farms served as a general model for countries that became socialist after World War II. In Eastern Europe and in China, state and cooperative farms more or less similar to those in the Soviet Union were organized. Except in Yugoslavia and Poland, they became the dominant form of agriculture by the early 1960s.

In the other socialist countries, there was less resistance to collectivization on the part of rich farmers than there had been in the

USSR because the changed balance of forces in the world made clear to many the impossibility of reversing the course of history and induced them to accept the change in order, with the possibility of living as prosperously as before, if with less power, working within a socialist framework. These countries were not under the same acute time pressure as the USSR had been to complete socialization in anticipation of a foreign invasion. Moreover, they started at a higher level of industrialization and, in the years immediately after World War II, benefited from Soviet assistance in provision of needed supplies and equipment. Consequently, the socialist governments of Eastern Europe were able to offer more material incentives to farmers to join cooperatives. Thus, the socialization of agriculture went more smoothly and, in general, did not pass through periods of large-scale sabotage and destruction by rich farmers.

Although there were mistakes, and in some cases premature organization of cooperatives which had to be disbanded, the Eastern European Communist-led governments won the support of the farmers. Thus, attempted anti-Communist revolts in Hungary and Czechoslovakia, originating in the cities, failed to win support in the countryside. This doomed the revolts to defeat, and contributed to the relative ease with which intervening troops from the Soviet Union and other countries put down the Hungarian revolt, and to the essential bloodlessness of the 1968 events in Czechoslovakia.

While the cooperative farms in Eastern Europe also went through stages of increasing socialization, there were variations in form and characteristics from country to country, depending on local conditions. For example, in the highly developed industrialized countries, the German Democratic Republic and Czechoslovakia, personal plots of farmers play less of a role than in the Soviet Union. On the other hand, a substantial number of private farmers remain in Romania, and a significant number in Hungary. In Poland, of course, the situation is decidedly different, and the limited socialized sector consists almost entirely of state enterprises. In most of these countries political parties of peasants cooperate with the leading Communist [arty, and have substantial representation in the governments, strengthening the cohesion between farmers and workers.

An example of extremely rapid increases in agricultural productivity under socialism is provided by the German Democratic Republic. The figures in Table 3.7 show the national income produced per worker in agriculture and forestry for the period 1950-74.

In 24 years, while the number of persons employed in agriculture declined by more than half, net agricultural production increased by about two-thirds. Productivity per person increased 3.7 times. The compound rate of increase was 5.6 percent for the entire period and 5.3 percent for the years 1965-74.

TABLE 3.7

National Income Produced per Worker in Agriculture and Forestry in the German Democratic Republic, 1950-74

Year	Persons Employed (thousands)	National Income Produced (billion marks, comparable prices)	National Income Produced Per Person Employed (marks)
1950	2,005	8,904	4,441
1955	1,721	10,993	6,388
1960	1,304	11,991	9,196
1965	1,179	12,147	10,303
1970	997	13,188	13,228
1974	903	14,776	16,363

Sources: Persons employed from Statistical Pocket Book of the German Democratic Republic (1975), p. 29. National income from Statistisches Jahrbuch der Deutschen Demokratischen Republik (1975), p. 36.

David J. Marcinko, while noting that the smoothness of transition to socialist forms contributed to the success of GDR agriculture, claims that gains in productivity were more marked while agriculture remained private, and declined after collectivization. This is in line with the general attempt of antisocialist economists to uphold capitalism as an intrinsically more efficient system than socialism.

Marcinko's argument is that 1960 was the year of maximum advance of collectivization, which was essentially completed in that year, and that thereafter, with collectivization completed, the rate of growth in productivity was considerably slower.[43]

However, a more careful analysis of the facts refutes his contention. Collectivization of agriculture was rapid throughout the second half of the 1950s. If socialized agriculture were intrinsically less efficient this would show up most conspicuously in its initial stages, during the early adjustment process, when social frictions resulting from the changed organizational form would be most harmful. Yet the rate of productivity growth during that five-year period did not decelerate.

True, 1960 was the year in which collectivization reached its maximum momentum. About 40 percent of all farms joined collectives and the process was essentially completed. By the middle of

1960, before the harvest, GDR agriculture was virtually 100 percent socialized. An exceptionally big increase in production was shown in 1960. Obviously this was accomplished by socialized agriculture and not by private agriculture, as improperly dealt with by Marcinko. Any reasonable time division would start the period of mainly socialized agriculture with 1960 or 1959, and the temporary slackening immediately thereafter, was obviously due to weather conditions--namely, an unusually good bumper-crop year in 1960--rather than to organizational factors.

True, the rate of productivity growth in the last decade, 5.3 percent per year, was a little lower than in the first decade of the German Democratic Republic. But in the earliest period, the gains were associated partly with recovery from wartime losses. In 1949-50, when the German Democratic Republic was formed, agricultural production was still significantly below the prewar level. At any rate, increases in productivity during the 1950s, both in the private sector and in the expanding socialist sector, took place in an economy already predominantly socialized, in which agriculture was supplied by industry under socialist planning, with centralized, planned incentives to agriculture, and with the farmers knowing that the government had a perspective of transforming all agriculture to socialist forms.

Currently about 10 percent of the gainfully employed people in the German Democratic Republic are supplying the population, with the aid of moderate imports, with a per capita food supply approximating that of the United States and other countries with high consumption standards. The social efficiency of GDR agriculture is on a par with that of Western Europe, and ahead of such a country as France, the archetypical stronghold of independent peasant and farmer agriculture.

Progress has been fully as remarkable in Bulgaria, which emerged from World War II a poor peasant country with low productivity. Bulgarian agriculture has certain advantages, however. It benefits from a relatively good climate and it escaped significant damage in World War II. Moreover, the Bulgarian Agrarian People's Union, with revolutionary traditions and a large following among farmers, was allied with the Communist Party in the revolutionary process, and remains an important factor in Bulgarian government and politics to this day.

Along with grain and livestock farming, Bulgaria specialized in fruit and vegetable growing. And Bulgarian industry developed equipment and chemicals suitable for this specialized farming, while Bulgarian science concentrated on it. Simultaneously Bulgaria developed a large-scale modern industry in particular areas of machinery and equipment, electronics, and other products. Taking maximum advantage of the opportunities for marketing throughout the socialist community provided by the CMEA, Bulgaria was able to produce particular agricultural and industrial products on a scale inconceivable on the basis of its domestic market alone.

By 1970, Bulgaria was able to project, and in the ensuing years carry out, the organization of the bulk of its agriculture into 160 agrarian-industrial complexes, which provide a more rational, planned use of all resources of land, labor, and capital than was possible on the basis of smaller and separate units of agriculture and industry. This process was watched closely by Soviet planners, serving for them as an experiment. It is now being applied, with suitable adaptations, in the Soviet Union, after first being used in Moldavia, which has natural conditions, within the Soviet Union, closest to those of Bulgaria.

The general forms of socialized agriculture pioneered by the USSR were followed also in the Asian socialist countries, with large variations corresponding to the wide difference from Soviet conditions in economic and social background. The most significant difference can be found in China. Even during the period of maximum Soviet-Chinese cooperation in industry, the Chinese leaders elected to go their own way in agriculture and accepted virtually no technical assistance from the USSR in agriculture.

In 1958-59, Chinese agriculture went over to the system of communes, ostensibly setups of a kind of primitive communism, in which all lived equally and communally, but at a very low level of life, owing to the low productivity and lack of mechanization of the countryside. Simultaneously, the communes were to become centers of local industry, such as the backyard iron and steel furnaces. However, except for brief experiments, the equal-distribution communes were not put into effect. The operative unit has remained the brigade, which is the former collective farm, and within it distribution is essentially according to the production of the individual farmer.

With all the social and political turmoil inside China, its socialized agriculture has performed better than that of similarly situated Asian countries which have remained capitalist. While food remains tightly rationed, the former mass malnutrition and frequent large-scale deaths from starvation have been overcome.

While statistics are sparse, visual evidence of the fields, the supplies in the stores, and the physical appearance of the population denote outstanding success for the agriculture of the Democratic Peoples Republic of Korea.

Cuban agriculture, since its socialist revolution, has had a distinct evolution. Because under capitalism the plantation system in sugar and latifundia system in cattle raising prevailed, with hired wage labor at the base, the main branches of agriculture were transformed into state farms instead of collectives. Meanwhile individual private farming has been retained in tobacco and other crops suitable for relatively efficient growing on smaller units.

During the 1970s, Cuban agriculture has developed close cooperation with that of other socialist countries within the framework of CMEA, which Cuba has joined. Because of its experience in fruit growing, Bulgarian assistance has been particularly important in Cuba in such areas as the design and provision of gardening equipment and the planning of the great citrus plantations being created on the Isle of Pines.

The success of socialized agriculture in the Soviet Union and other countries has contributed to the inspiration of movements for redistribution of large landed estates to those who till them, and for formation of production cooperatives, in countries where capitalist and even precapitalist agricultural forms still prevail. Examples are the land reform and formation of cooperatives in Chile during the Allende regime, and more recently in Portugal. A similar process has been underway in Peru. In Guyana, the large sugar plantations have recently been nationalized and cooperatives formed among the small farmers.

However, the breakup of cooperatives in Chile after the counterrevolutionary coup, their struggle for survival in Portugal, the failure of the ejido movement in Mexico, and the dwindling of the kibbutz movement in Israel, illustrate the point that socialized agriculture cannot flourish in a predominantly capitalist environment. Just as labor has to be the leading force politically in building socialism, so industry has to be socialized and the overall economy planned to provide the necessary political and economic foundation for the successful socialization of agriculture.

Given that foundation, the overall record, in the Soviet Union and elsewhere, establishes the economic superiority of socialism over capitalism, in agriculture as in industry.

PROBLEMS AND PROSPECTS

The Soviet government and the collective farms have poured tremendous resources into agriculture on a rapidly expanding scale since World War II.

Capital investments in agriculture increased from about 15 billion rubles in the five-year period 1951-55, to 48 billion in 1961-65, 132 billion in 1971-75, and a planned 172 billion rubles in 1976-80. These were or are planned to be, respectively, 16, 20, 26, and 28 percent of total investments.[44]

Some of the increase from the early 1950s to the early 1960s represents a broadening of the coverage of investments attributed to agriculture, and some of the increase represents rising costs of capital construction and equipment, but most of the increase is real. The scale of the effort is really impressive. Annual Soviet investments in agriculture are currently considerably larger than U.S. investments in agriculture, even with any reasonable adjustment for differences in definition.

The investment has been particularly heavy in farm equipment and vehicles of all kinds, in premises and equipment for automated livestock raising, and for irrigation and drainage projects. A substantial and increasing part of the investments have been for "nonproductive" purposes; that is, to provide the farmers with modern housing and community facilities.

In addition, there has been a very rapid increase in the supply to agriculture of fertilizers, pesticides, and herbicides. Economic incentives to farmers have been improved in various ways so that their per capita income has gone up more rapidly than that of workers in any other industry, despite the slower growth of farm production, as compared with industrial production.

Yet the increase in agricultural output has been consistently less than planned, and has obviously fallen far short of keeping pace with the increased resources put at the disposal of agriculture. This is dramatically illustrated by the following figures: between 1960 and 1974 the gross production of Soviet agriculture increased 50 percent, while the national income produced in Soviet agriculture increased only 6 percent.[45]

Thus, most of the growth in output represented increased inputs, including increased depreciation on the larger stock of capital equipment. The net value added created by Soviet agriculture increased very little over that 14-year period.

A Soviet economist, V. Popov, found the increase in output per 1 percent increase in capital in place in Soviet agriculture equal to only 0.32 percent in 1961-65, to 0.66 percent in 1966-70, to 0.27 percent in 1971-75, and with a planned ratio of 0.46 to 0.51 percent for 1976-80. Capital invested in industry proved much more effective in increasing production.[46]

Certainly the weather has had much to do with the effectiveness of Soviet investments in agriculture. The relatively good results in 1966-70 were associated with reasonably normal weather, for the USSR, while the 1970s included three years of bad to very bad weather for agriculture. A large part of the Soviet investments are devoted to reducing the impact of adverse weather and climate. The most important are the irrigation and drainage programs, which aim to provide sure crops in drought-prone areas and to make swampy areas with adequate rainfall cultivatable. Construction of adequate winter quarters for livestock and storage facilities for newly harvested grain aims to avoid losses caused by freezing of animals and soaking of crops.

But there are very serious social and economic problems which substantially reduce output and productivity in all weather conditions. And no Western critics are more vigorous than the Soviet economists in exposing these problems and proposing solutions for them.

One of the most painful problems is demographic, largely a consequence of World War II, and even more remotely, of the Civil War following the revolution. As a result of the earlier conflict, and subsequent lesser struggles, by 1940 there were 8 million more women than men in the Soviet Union. After World War II the discrepancy increased to 22 million, and has only slowly diminished to 18 million in 1976. The entire discrepancy is in people over 45 years old. And it is heavily concentrated in the countryside and especially on the collective farms. Most of the soldiers in World War II came from the farms, and deaths among them left millions of widows and never-married women on the collective farms. Moreover, after the war young men completing their military service would tend to take jobs in industry, while the young women had less opportunity, or venturesomeness, to leave. Thus many collective farms today have not only a majority of women workers, but predominantly older women, with limited education and without specialized training.

In a capitalist society these women might be forced to exist on charity or "welfare." Under socialism, they have the right to employment. Many of them can be seen on the collective farms doing auxiliary work--hand-picking potatoes or sugar beets, gathering stray sheaves of grain behind the combine, caring for livestock without mechanical aids.

Their employment on the collective farm adds to the total production, but reduces the overall indexes of labor productivity. Moreover, payment systems require that the less productive workers receive a minimum sufficient to meet essential needs. This reduces the funds available to stimulate a high level of efficiency and stability of employment among the machine operators and specially trained workers, men and women, on whom the bulk of production depends.

Soon all the war widows will be retired--the standard retirement age for women in the USSR is 55 years. Moreover, reversing earlier trends young women are leaving the farms more rapidly than young men, because they have fewer opportunities at rewarding jobs on the farm and ample opportunities in office, service, and factory jobs in the cities.

Soviet writers claim that the social environment on the farm does not attract skilled workers. Recent improvements in the form of building larger, urban-type settlements, club houses, and cinemas have not been sufficient. The young people especially want access to the full life of urban centers, while many of them are lured by the great construction projects, often in remote parts of the country. A young man, trained as a tractor driver, can turn that skill to operating a bulldozer on a construction project. Between 1961 and 1972, 10 million agricultural equipment operators were trained, but the number employed in agriculture increased only a little more than one million.[47]

Because the people getting a higher level of education can get and prefer good jobs off the farm, the average educational level of farm workers, even those employed as engineers and in other technical capacities, is lower than that of their counterparts in nonfarm occupations, and considerably below the level needed for scientific agricultural practices.

The increase in mechanization has been steady, but still not enough to eliminate the need for a great deal of hand labor. For example, by 1974 the level of mechanization had only reached 24 percent in feeding livestock, and the level of complex mechanization had reached 21 percent in cattle raising, 38 percent in raising swine, and 49 percent in poultry farming.[48] This results in a large volume of unproductive, heavy, and unpleasant hand labor, increases production costs, and complicates problems of keeping energetic, skilled labor.

Many Soviet writers note that the existing level of mechanization would go much further if the equipment were used on a two-shift basis. Efficiency in milking of cows is very low, for example. There are more than 3 million milkmaids in the USSR, caring for an average of a little more than 15 cows each. Moreover, generally they have to work a triple split shift, for the three daily milkings.

This in turn leads to the high cost of dairy farming, and the unprofitability of this branch of agriculture, in comparison with crop raising, even after dairy and livestock prices were increased substantially in 1970. The result is that farmers have insufficient incentive for the development of livestock production. Actually it is increasing at a priority rate, but not as much as Soviet planners would like, and apparently not commensurate with the share of added resources provided for it.

Along with the departure of the better workers from farms in many parts of the country, there is an unneeded increase in the farm population in some of the southern republics, notably some of the Transcaucasian and middle Asian republics. In the early 1970s, productivity of labor in socialized agriculture was only 39 percent of the national average in the Georgian Republic but 185 percent in Estonia.[49] Increases in labor productivity have been very fast in the Baltic republics, relatively slow in Georgia and Uzbekistan.

With the favorable climatic conditions in Georgia, people can live well concentrating on their personal plots, and doing a minimum amount of work on the cooperative or state farm. Thus, in Georgia in 1976, only 44 percent of the cattle and 39 percent of the pigs were owned by the collective and state farms, as compared with 79 percent and 62 percent respectively in Byelorussia, and 83 percent and 93 percent in Estonia.[50]

Productivity of milk per cow and eggs per chicken is much less on the personal plots than in the specialized socialist dairy and egg enterprises, and the consumption of feed per pound of product runs twice as high.

Soviet self-criticism of its agriculture is generally constructive and optimistic in tone. The existing shortcomings have to be evaluated in the light of the history of Soviet agriculture. While progress has fallen short of plans, it has exceeded any pace of advance achieved by capitalist agriculture, and it has fulfilled its essential tasks within the overall progress of Soviet life.

But the lagging of agriculture behind plans has somewhat held back the pace of growth in economic power and living standards. For two decades, Soviet political and economic leaders have taken a succession of initiatives aimed at improving the overall performance of agriculture. Most of these have made a contribution--the opening up of virgin lands in Kazakhstan and the Altai, the raising of procurement prices, the regularization of collective farm payments and inclusion of collective farmers within the system of social security, the merging of collective farms and increase in the scope of state farms, the priority in provision of increased trading and service facilities to the countryside, and of incomes to the farmers, the huge investments in irrigation, land melioration, agricultural chemicals and machinery, and the increased independence of the collective farms in working out their production and investment programs.

Special social interest attaches to the program adopted in 1975 to improve the opportunities and conditions for women to work as tractor drivers and other farm machine operators.

Thus, while only 7,000 women drivers were working in the fields of the Russian Federation (and 3,000 in the Ukraine), more than 56,000 were taking courses as tractor drivers and other farm equipment operators in schools and institutes of the Russian Federation.

Under the program, women tractor drivers are given the most comfortable, up-to-date machines, they are assigned work norms

10 percent below those of men on the same farms, they get six days extra annual holiday, and are eligible for retirement after 15 years of work at the age of 50.

This, indeed, is "affirmative action"![51]

At this stage, one main emphasis is on the acceleration of creating large, modern, specialized farming enterprises and industrial enterprises related to agriculture within the framework of interfarm complexes. Another emphasis is on creating conditions for highly productive agriculture in the "non-black-earth" zone, of Central and Northern European Russia, where agriculture traditionally has been low-yielding owing to swampy, acid soil. Billions of rubles are being invested in drainage, melioration, chemical treatment of the soil, and in erection of facilities in this area, for multiplied output of foodstuffs for the population in the main industrial and cultural centers of the country. The long-considered proposal to reverse the flow of northern rivers for irrigation and other purposes is finally approaching the practical design stage, while the pace of irrigation works in the drought-prone areas continues at a very high level.

The private plots will continue to play an important role in Soviet agriculture as long as there are large numbers of rural people with a peasant background who like gardening and feel the need to supplement their incomes from socialized agriculture, and as long as the marketing and transportation network is not sufficiently developed to supply the rural population with all of their needs in foodstuffs conveniently, and with nonfood commodities sufficiently attractive and inexpensive to serve as an incentive for maximum efforts in socialized farming for the market. Official doctrine stresses the importance of the private plot in the provisioning of the Soviet people. But many Soviet economists look forward to its ultimate elimination.

V. Nikiforov points out the "contradictory" character of the private plot, from the point of view of the farmer. On the one hand, it helps equalize farm and city incomes. On the other hand, it engenders economic problems associated with private enterprise, it "increases the length of the workday and reduces the amount of free time that is necessary for the comprehensive development of people." He concludes:

> The solution of all these problems is possible only with the extinction of private subsidiary farming through the development of social agricultural production and the growth of income . . . which will make it possible to compensate both society and individual groups in the work force for the curtailment of production in private farming.

But he cautions,

> At present we must not underestimate its role as an additional source of agricultural production and of increasing the incomes of the working peoples, which must be used by virtue of the insufficient development of social production.[52]

The process of drawing together of collective and state farm forms has accelerated in recent years, and the formation of interfarm complexes will speed this process. Despite official statements to the contrary, the transformation of collective farms into state enterprises, as the democratic, cooperative character of the collectives is extended increasingly also to the state farms, seems a not unlikely development in coming decades.

A Soviet sociologist, P. Simush, with a group of associates, during the middle 1960s intensively studied the life of the Rossiya Collective Farm in the Kuban Cossack area north of the Caucasus mountains. He recounts everything about the history of the farm, its difficulties, the mistakes of its managers, the wartime tragedies. He also explains the steps by which gradual improvements were achieved, how the farm achieved stability and a relatively good standard of living and quality of life for its inhabitants. A poll of 305 collective farmers found that 68 liked their work very much, 164 were satisfied with it, 11 were indifferent, 16 were completely dissatisfied, and 46 had no opinion.

Of the same sample, 200 would not change their job, 44 would, and 61 did not know. Many spoke eloquently of their feeling of participation, their joy in creative labor. Many were studying to raise their qualifications.[53] This was a rather advanced farm ten years ago. But much of Soviet agriculture has caught up to that level, or gone beyond it.

In presenting the final version of the five-year plan for 1976-80, and the plan for the year 1977, State Planning Commission Chairman N. K. Baibakov said that with fulfillment of this five-year plan, Soviet agricultural production would increase from 85 percent of the U.S. level in 1975, to 100 percent of the 1975 level of U.S. agricultural production in 1980.[54]

Official U.S. estimates agree with Baibakov, approximately, on the recent ratio of Soviet to U.S. farm output.[55] Even when it reaches 100 percent of the 1975 level of the United States, or even 100 percent of the then current U.S. level, Soviet agriculture will be well behind the U.S. level in production per capita of the population, and in labor productivity. But it may be well ahead of U.S. agriculture in social conditions, in the well-being of the rural population. And at any rate, the agriculture of the Soviet Union and other socialist countries is already important in the historic economic competition between socialism and capitalism.

Meanwhile, socialist agriculture is an increasing element in economic and technical cooperation between the two social systems, including between the United States and the Soviet Union. The sale of grain by the United States to socialist countries, the purchase of meat from socialist countries by the United States, are current examples, which will certainly be expanded as existing U.S. restrictions on trade with socialist countries, and embargoes on trade with some, are lifted.

NOTES

1. Central Statistical Administration, Selskoye Khoziaistvo, Statisticheskii Sbornik (Moscow, 1960), p. 399.

2. Institute of Geography, USSR Academy of Sciences, Man, Society and the Environment (Moscow: Progress Publishers, 1975), p. 283; Narodnoe khoziaistvo SSSR v 1964 godu (Moscow, 1965), p. 38.

3. Narodnoe khoziaistvo SSSR v 1964 godu (Moscow, 1965), pp. 32, 33; Narodnoe khoziaistvo SSSR v 1974 godu (Moscow, 1975), pp. 7, 448, 548; SSSR v tsifrakh v 1975 godu (Moscow, 1976), p. 171.

4. U.S. Department of Commerce, Historical Statistics of the United States, Colonial Times to 1970, Series D 5-7, K 502-16 (Washington, D.C.: U.S. Government Printing Office), pp. 139, 167-215.

5. Quoted in Maurice Dobb, Soviet Economic Development Since 1917 (New York: International, 1968), p. 222.

6. U.S. Department of Commerce, Bureau of the Census, Statistical Abstract of the United States, 1975 (Washington, D.C.: Government Printing Office, 1975), no. 1062, p. 628; no. 1066, p. 630.

7. Narodnoe khoziaistvo SSSR v 1974 godu, p. 357. The statistics refer to winter wheat; but only relatively trivial acreages are planted to spring wheat in these Union Republics.

8. Marx, Engels, Lenin, On Communist Society (Moscow: Progress, 1975), p. 19.

9. Lenin, Collected Works, vol. 29 (Moscow: Progress Publishers, 1965), pp. 419-23, 427-31.

10. Ibid., vol. 32, pp. 214-17.

11. L. Nikiforov in Voprosi ekonomiki, February 1975, translated in Problems of Economics, November 1975, p. 78.

12. Statistisches Jahrbuch der DDR (Berlin: Staatsverlag der Deutschen Demokratischen Republic, 1975), 22: 70.

13. U.S. Department of Agriculture, Farm Labor (Washington, D.C.: USDA, August 26, 1976); U.S. Department of Labor, Employment and Earnings (Washington, D.C.: Government Printing Office).

14. U.S. Department of Commerce, Survey of Current Business (Washington, D.C.: Government Printing Office, July 1976), Tables 6.6 and 6.10, pp. 51-52.

15. Narodnoe khoziaistvo SSSR v 1974 godu, p. 606.

16. Ibid., pp. 422-23.

17. Ibid., pp. 422, 443.

18. A. Emelianov, "The Party's Agrarian Policy and Structural Changes in Agriculture," Voprosi ekonomiki, no. 3 (1974), translated in Problems of Economics, November 1975, pp. 104-05.

19. A. N. Kosygin, "Guidelines for the Development of the National Economy of the USSR for 1976-1980," in Socialism, Theory and Practice, April 1976, p. 18.

20. Nikiforov, op. cit., p. 71.

21. Pyotr Simush, The Soviet Collective Farm (Moscow, 1971), p. 132.

22. The Development of the Czechoslovak Economy, materials of a symposium, Prague, April-May 1973, Appendix, Table 31, p. 191.

23. U.S. Department of Commerce, Consumer Income, Money Income in 1974 of Families and Persons in the United States, no. 101 (Washington, D.C.: Government Printing Office), p. 60, Table 64, p. 131.

24. Gertrude E. Schroeder and Barbara S. Severin, "Soviet Consumption and Income Policies in Perspective," Joint Economic Committee, U.S. Congress, Soviet Economy in a New Perspective (Washington, D.C.: Government Printing Office, 1976), pp. 622-23.

25. Pravda, July 24, 1976.

26. Historical Statistics, Series G 881-915, pp. 329, 331; and U.S. Department of Agriculture, National Food Situation, November 1976, p. 15.

27. Statistical Abstract of the United States, 1975, no. 149, p. 92.

28. New York Times, January 5, 1977, p. A3.

29. Daily World, December 16, 1976.

30. People's Republic of Bulgaria, Central Statistical Administration, Statistical Reference Book, 1976, p. 88; and Gancho Krustev in Central Committee of the Bulgarian Communist Party, Information Bulletin, no. 10, 1976, p. 61.

31. Central Intelligence Agency, Handbook of Economic Statistics 1976 (Washington, D.C.: CIA, September 1976), Table 1, pp. 18-19.

32. SSSR v tsifrakh v 1975 godu, p. 202.

33. Statisticheskiye Ezhegodnik, Stran-Chlenov Sovieta Ekonomicheskoye Vzaimopomoshi, 1975, "Statistika" (Moscow, 1975), no. 4, pp. 11-12; no. 69, p. 204.

34. Ibid., no. 4, pp. 11-12; no. 24, p. 51. Romanian consumption estimated from production given in no. 69; and exports given in Anuarul Statistic al Republicii Socialiste Romania (Bucharest: Directia Centrala de Statistica, 1976), no. 187, p. 393.

35. Wall Street Journal, September 15, 1976.

36. Statisticheskiye Ezhegodnik, op. cit., no. 117, pp. 395-96; no. 17, p. 42.

37. V. Matskevich, Agriculture (Moscow: Novosti Press Agency Publishing House, 1971), pp. 40-42.

38. Institute of Marxism-Leninism, Great Patriotic War of the Soviet Union, 1941-45 (Moscow, 1974), pp. 141, 341.

39. G. Deborin, Secrets of the Second World War (Moscow: Progress Publishers, 1972), pp. 141-42.

40. Great Patriotic War, pp. 141, 342.

41. Deborin, op. cit., p. 133.

42. Great Patriotic War, p. 142.

43. David J. Marcinko, "Total Factor Productivity and Collective Agriculture in East Germany," Association for Comparative Economic Systems Bulletin, Winter 1976, pp. 49-58.

44. Narodnoe khoziaistvo v 1974 godu, pp. 528-29, SSSR v tsifrakh v 1975, pp. 161, 163, Planovoe khoziaistvo, no. 9 (1976), editorial p. 9; Pravda, October 28, 1976.

45. Statisticheskiye Ezhegodnik, op. cit., no. 14, p. 40; no. 58, p. 165.

46. V. Popov, "Povuishenniye Proizvoditel 'nosti Truda v Sel'skom Khoziaistve," in Voprosi ekonomiki, no. 8 (1976), p. 83.

47. V. Kostakov, "Economical Utilization of the Labor of the Employed Population," Planovoe khoziaistvo, no. 7 (1974), translated in Problems of Economics, May 1975, p. 82.

48. Popov, op. cit., p. 85.

49. Ibid., p. 87.

50. SSSR i Soyuziye Respubliki v 1975 godu (Moscow, 1976), pp. 85, 141, 315.

51. M. Fedorova, "Ispol'zovanie Zhenskovo Truda v Sel'skom Khozyaistve," in Voprosi Ekonomik, no. 12 (1975).

52. Nikiforov, op. cit., pp. 78-79.

53. Simush, op. cit., pp. 63-68.

54. Pravda, October 28, 1976.

55. David W. Carey, "Soviet Agriculture: Recent Performance and Future Plans," Joint Economic Committee, U.S. Congress, Soviet Economy in a New Perspective (Washington, D.C.: Government Printing Office, October 14, 1976), p. 577, gives a ratio of Soviet to U.S. agricultural output of "about four-fifths" in the early 1970s.

CHAPTER

4

THE SOCIALIST TRANSFORMATION OF AGRICULTURE

Novosti Press Agency

Soviet agriculture is a major branch of the Soviet economy bringing in one-third of the USSR's national revenue and meeting 75 per cent of the demand for consumer goods. . . .

Though the territory of the Soviet Union is vast not all of it is suitable for agriculture. One-third of it is forested; there are large regions which are tundra (treeless plains), mountainous areas, marshland and deserts. Thus, less than a quarter of the country's entire territory--548 million hectares (224 million hectares are ploughland) can be used for agricultural production.

Climatic conditions in the USSR are highly diversified and so are the conditions of farming. Sixty per cent of ploughland is in an area with difficult farming conditions (the US has only one per cent).

The biggest trouble for farmers in the Baltic region, Byelorussia and the center of the Russian Federation is excessive soil moisture. To combat this, drainage is being effected on a wide scale.

Compiled and reprinted by permission of the Novosti Press Agency, Moscow, and of the Soviet Embassy in Washington, D.C., from USSR Agriculture (Moscow: Novosti Press Agency Publishing House, 1973), pp. 6, 8, 10, 13, 14, 18, 50, 55-57, and 110; USSR Agriculture (Moscow: Novosti Press Agency Publishing House, 1975), pp. 4, 5, 12, 13, 18, 20, 21, 50, 55-57, 96, 101, 103-05, 110, and 111; USSR 75, Novosti Press Agency Year Book (Moscow: Novosti Press Agency Publishing House, 1974), pp. 195-200; Leo Voskerensky, Novosti Agency Economic Observer, "Farming: Plans to Beat the Weather," Soviet Life, June 1976, p. 10.

And conversely, the Ukraine, the Volga areas and Kazakhstan--the country's major grain producers--suffer from lack of moisture. Once in three or four years, or even more frequently, drought strikes extensive areas, at times all the way from the Altai to the Carpathians, and causes damage to agriculture.

Large irrigation projects are being built to bring water to drought-affected lands. However, the solution of the problem as part of the comprehensive programme of agricultural development takes time and vast capital investments as well as material and technical resources. . . .

THE ORGANIZATION OF AGRICULTURAL PRODUCTION

Land in the USSR is national, state property. It is neither bought nor sold. The bulk of the farmland is used by the collective farms (kolkhozes) and by state farms (state-owned agricultural enterprises). Collective farms have 106.8 million hectares at their disposal, or about half of the 224 million hectares of the country's sown area.

Part of the land is used by various state experimental farms, poultry and stud farms, educational and research institutions.

Besides, collective farmers and state farm employees are given small household plots of land (about half a hectare) from the stocks of the collective or state farms. These household plots are used as vegetable gardens, orchards, or vineyards. Plots of this kind account for 7.8 million hectares (3.5 per cent of the total).

Collective Farms are the voluntary agricultural co-operatives of farm workers. The state grants the collective farms the land in free and permanent use and provides them with modern equipment, chemicals and, whenever necessary, loans for seeds and bank credits. Moreover, the state is the chief wholesale buyer of the products of the collective farms, guaranteeing them a permanent market.

Collective-farm property--farm buildings and machines, livestock and the output--belongs to all member-farmers. They are the sole owners of the fruits of their labour. The supreme administrative body of the collective farm is the general meeting of all members, which elects the management board and the chairman. The general meeting endorses the collective farm's development plans and the board's annual reports, and makes all major decisions concerning the life and activity of the farm. In all their work the chairman and the board members are accountable to the general meeting.

Members of collective farms are remunerated in accordance with the quantity and quality of their labour. In the last few years collective farms have switched over to guaranteed monthly payment.

The collective farm of today is a large mechanized economic unit. On the average, it has about 6,300 hectares of land, 3,200 hectares of it ploughland; over 4,000 head of livestock (cows, sheep, pigs), 34 tractors, 10 harvesting combines and 17 lorries.

In 1973, the country had 31,500 collective farms.

Collective farms provide roughly 50 per cent of the total commercial products of the country's agriculture.

They receive the bulk of their income from the sale of their goods to the state at stable (zonally differentiated) prices and according to stable plans of deliveries. This assures steady economic incentives for agricultural production.

Surplus products may be sold on the urban market either independently or through the rural consumer co-operative "Tsentrosoyuz" (Central Union of Consumer Societies of the USSR).

State Farms are state-owned agricultural enterprises. The state farm is supervised by a director appointed by the relevant state body. The state farm employees are united in a trade union, which, as in every socialist enterprise, plays a considerable role in the management of production. They draw fixed wages, but in addition, like other agricultural workers, they have a plot of land for their own use.

There are over 16,000 state farms in the country. These are large, highly mechanized enterprises with a sound economic base for effective agricultural production. The average state farm has 20,000 hectares of land, 6,700 hectares of it ploughland; over 7,000 head of livestock (cows, sheep, pigs), 55 tractors, 20 harvesting combines, and 26 lorries. Most state farms are specialized.

The state farms produce approximately half of the country's commercial grain and vegetables, a fifth of the raw cotton and more than forty per cent of the animal products. All their commercial produce is purchased by the state.

Personal Plots. The household plots of collective farmers and state-farm employees still account for a big percentage of the total production of vegetables, milk and especially eggs.

Most of the products of the household plots are consumed by the owners themselves; the rest is sold either on the free market, or through the consumer co-operative organization "Tsentrosoyuz."

In early 1974, the personal property of farmers comprised 14.6 million cows, 13.5 million pigs, 37.4 million sheep and goats, and hundreds of millions head of poultry.

Agricultural Management. The guidelines for agriculture are determined by the Ministry of Agriculture of the USSR.

The Union Republics have their own ministries and regional and district departments of agriculture.

The USSR Ministry of Agriculture deals with the basic planning and economic problems of agricultural development throughout the country. It co-ordinates the plans of agricultural produce output, promotes the development of agricultural science and the generalization and dissemination of the best methods of agricultural production, and sees to the provision of agriculture with modern equipment, chemical fertilizer, chemical and biological pest-killers, and to other related matters.

Prompt assistance and scientifically-based recommendations on managing agricultural production are given by the local regional, and especially the district, agricultural departments, which are directly connected with the collective and state farms.

The total number of specialists employed in agriculture exceeds 950,000. Eighty-eight per cent of the collective farm chairmen and almost all state farm managers have received specialized education.

THE WAY IT WAS

In pre-revolutionary Russia the peasants comprised over three-quarters of the population. Agriculture was extremely backward and productivity was low. Almost two-thirds of the land, and the best land at that, belonged to the landowners, the tsar, the church, and the kulaks--the village rich. Of the 20-odd million peasant holdings 13 million were owned by poor farmers who eked out a miserable and wretched existence, and were mercilessly exploited by the landowners and the kulaks. Every year two million peasants migrated to the southern part of the country to earn a living. There they were hired by the kulaks and the rich landowners who took advantage of their desperate situation and made them work 12-14 hours a day for a mere pittance. The predominant mass of the peasantry worked the land with primitive farm implements--11.2 million wooden ploughs and about 18 million wooden harrows. According to statistics for the end of 1916, only 0.8 per cent of the farm implements were mechanised, and draught cattle accounted for 99.2 per cent of the power used on the farms. Of the total number of farms in pre-revolutionary Russia, in 15 per cent the farmers did not own the fields, 30 per cent had no horses, and 34 per cent were completely lacking in farm implements of any kind.

The poor peasants could neither buy nor use farm machinery or more efficient farm implements, nor could they use more progressive methods of farming, such as multiple crop rotation.

The general backwardness in agriculture manifested itself in the extensive mode of farming and its monocultural character. For example, grain crops, mostly rye, barley and oats, were grown on 90 per cent of the arable land leaving less than 10 per cent of the tilled land for technical crops, potatoes, vegetables, melons, watermelons and forage crops.

The backwardness and one-sidedness of agricultural development had an adverse effect on livestock breeding.

The cattle was mostly put out to pasture and fed hay, straw and chaff. Therefore the size of herds and cattle productivity grew very slowly, and the per capita output of livestock products was extremely low. Most poor farmers' families very rarely tasted milk or meat. Famine, epidemics, and poverty were the lot of the Russian peasants.

Such was the situation in the Russian countryside before the Great October Socialist Revolution.

TOWARD THE SOCIALIST TRANSFORMATION OF AGRICULTURE

The Decree on Land adopted by the Soviet Government right after the victory of the Revolution reflected the most vital interests of the toiling peasantry. In the first place it formalized the nationalization of land and stipulated the principles of the Soviet government's land policy. The decree proclaimed the following:

1. Landed proprietorship was abolished forthwith without any compensation.

2. The landed estates and also all crown, monastery, and church lands, with all their livestock, implements, buildings and everything pertaining thereto had to be placed at the disposal of the land committees, and the Soviets of Peasants' Deputies pending the convocation of the Constituent Assembly.

The decree pointed out that the confiscated property belonged to the people and stressed the need for public protection of this property. It did away with private ownership of the land stating:

"Private ownership of land shall be abolished forever; land shall not be sold, purchased, leased, mortgaged or otherwise alienated."

For the first time in history the victory of the Great October Socialist Revolution realized the peasants' age-old dream of land, and rooted out the remnants of serfdom in agriculture, thereby creating favourable conditions for the gradual transition to socialism in agriculture.

More than 150 million hectares of land which had once belonged to big landowners, the czar, monasteries and the government were handed over to the peasants by the Soviet Government without compensation. The peasants were freed from the payment of annual rents and from expenditures on buying the land they tilled, as well as from all their debts to the Peasant Land Bank. The implements which had belonged to the landowners, worth over 300 million roubles in all, were given to the peasants. . . .

The victory of the Great October Socialist Revolution led to a substantial regrouping of the class forces in the country. The bourgeoisie and landowners as a class had disappeared. The economic might of the rich peasants greatly weakened. A large number of the poor peasants were coming closer to becoming middle peasants after receiving plots of land, implements and cattle previously owned by rich landowners and kulaks. Whereas the pre-revolutionary period gave rise to large capitalist farms and led to the ruin of small landowners, after the revolution the middle peasant became the central figure in the countryside.

It was necessary in those conditions to form the right sort of economic and political relations between the working class and the peasantry, and to charter ways for the gradual transformation of the small-commodity farms into large-scale socialist production. . . .

The transformation of farming along socialist lines is the most difficult and complicated task. . . . Whereas it was comparatively easy to take away the large-scale capitalist industrial plants in the towns from the bourgeoisie and turn them into national property, it was quite a different matter in the countryside, for there the Soviet Government had to deal with small peasant households.

Here socialism had to be attained in a different way, which required both time and preparation. In the first place it was necessary to build a material and technical base, to strengthen the alliance between the working class and the toiling peasants in their struggle against rich peasants, to conduct wide-scale political and cultural work among the masses in the countryside, to change their psychology, to remould the entire make-up of village life along socialist lines, to convince the peasants of the advantages of large-scale socialist production, using the experience of the first collective and state-owned farms as an example.

AGRICULTURE IN WAR AND IN PEACE: THE SOVIET WAY

The movement for the collectivisation of agriculture was started by the village poor. The Soviet state supported it and helped the first collective farms in every way possible: it granted credits, extended subsidies, supplied tractors and other farm machinery. An important step in this direction was the setting up of machine-and-tractor stations which served as a stronghold of socialism in the village. It took some time, however, before the peasantry, as a whole, felt it was advantageous to join a collective. . . .

Collectivisation put an end for all time to the exploitation of the rural masses by the village bourgeoisie, and to class antagonism in the countryside. The establishment of public ownership of the means of production on the farms created favourable conditions for better organisation of agricultural production and for better utilisation of land.

From a backward, agrarian country, which Russia was before the Revolution, the Soviet Union developed into a nation with large-scale progressive farming based on scientific achievement and the wide use of machinery.

Before the Socialist Revolution of October, 1917, the peasants, as we mentioned earlier, tilled their land with very primitive instruments--wooden ploughs and harrows, etc. Twenty-three years later (seven of them were the years of the First World War and the Civil War), 62 per cent of the spring crop ploughing was done by tractors, 56 per cent of the sowing was done by tractor-drawn sowing machines, and 42 per cent of the grain was gathered by harvesters.

The social make-up of the village changed beyond recognition. New trades and occupations became popular. By 1940 there were 1.5 million tractor drivers, harvester operators, lorry drivers and engineers. At present 33 per cent of the farmers handle machinery.

As the collective farm movement gained scope in the countryside, a cultural revolution was taking place. This could be seen from the appearance of specialists with a higher or secondary special education; agronomists, livestock experts, and veterinarians. There were about 50 thousand such specialists at the collective and state farms and at machine-and-tractor stations by early 1941.

Great changes took place in the cultural life of the Soviet village. Gone were the days when the majority of the population could not read or write. Illiteracy, that harsh legacy of the past, had been completely wiped out before the war (by 1941). The collective farmers willingly attended ABC courses and their children received a secondary or even a higher education.

The collective farm system based on the implementation of Lenin's cooperative plan has stood the test of time; its vitality was vividly demonstrated during the Great Patriotic War (1941-45). Despite the military occupation of important agricultural areas by the enemy, and the acute shortage of a skilled work force, the collective farmers throughout the country went all out in their efforts to provide the Soviet Army and the rear with foodstuffs, and industry with raw material. Therein was revealed the strength of the new social relations in the village, the true patriotism and the very real sense of duty of the collective farmers.

Agriculture suffered greatly during the war. The Nazi invaders destroyed or plundered more than 70 thousand villages, 98 thousand collective farms, 1,876 state farms and 2,890 machine-and-tractor stations. Despite all these difficulties the collective farmers and state farm workers rebuilt their farms and put them back into working condition in record time. By 1950 the overall output of the Soviet collective and state farms had reached the pre-war level.

In later years, however, the rate of development of this important branch of the national economy slowed down, creating some interruptions in the supply of foodstuffs and raw material for light industry. This was, to a certain extent, due to financial and technical difficulties in the first few post-war years. Also the most was not being made of the possibilities that the socialist mode of production at collective and state farms had to offer.

The Soviet Communist Party and the Soviet Government took all the necessary measures to help the Soviet farmers with machinery and equipment and give them added incentive to work. An important factor in the growth of farm production was the pioneering and development of the virgin and fallow lands on the vast plains of Kazakhstan, Siberia, the Volga area and other parts of the country where more than 40 million hectares of land were put to the plough. Hundreds of new 25-30 thousand hectare state farms have sprung up there. The newly developed virgin and fallow lands have become an important source of grain and other agricultural produce.

The implementation of the Eighth Five-Year Plan (1966-70) has wrought great changes in the life of the farmers. It has strengthened the material and technical base of the collective and state farms which are now getting far more farm machinery, fertilizers and other chemicals than before. Great progress was made in bringing electric power to the villages, and in improving soil conditions. As a result, the average annual output of agricultural produce is now increasing at the rate of 21 per cent a year compared with 12 per cent in the previous five-year plan period. The most spectacular results were achieved in grain production. The average grain harvest increased 37 million tons, or about 30 per cent. All the other branches of farming also made marked headway.

The state allocated about 129 thousand million roubles for the further development and intensification of agriculture in the ninth five-year plan period. This is approximately the total amount invested in agriculture for the last ten years.

SOVIET AGRICULTURE IN THE 1970S: ACHIEVEMENTS AND PROSPECTS

Over the past five-year period (1971-1975) the country's average annual volume of farm production grew by 13 per cent. The next five-year period envisages a 14-17 per cent rise while population growth remains at the five per cent level. Thus, the production of food and agricultural raw materials for industry will continue to outstrip population growth.

This creates favorable conditions for the main goal of the five-year plan: a steady rise in the standard of living.

Estimates indicate that the production of about one metric ton of grain per capita would meet all the country's needs. That would require a gross yield of more than 250 million tons. But such a target would exceed all possible resources, and so the Tenth Five-Year Plan has a more moderate but more practicable goal: 215-220 million tons.

The consumption of bread and bakery products goes down from year to year (per capita consumption was 164 kilograms in 1960, 156 kilograms in 1965, 149 kilograms in 1970, and 147 kilograms in 1975), but the average annual per capita production of grain is growing with equal consistency. In 1961-1965 it was 561 kilograms; in 1966-1970 it was 687 kilograms, and in 1971-1975 it was 711 kilograms. The logic is clear: The higher grain yields make possible more intensive livestock breeding and, as a consequence, increased consumption of meat, milk and eggs. That is why the grain harvest is the key index of the agricultural program.

With an annual yield of 215-220 million tons of grain, the Soviet Union can keep improving the diet of its citizens.

Grain Yield Grows Despite Bad Weather Conditions

The natural conditions for farming in the USSR are not nearly as favorable as in any of the advanced Western agricultural countries. On 40 per cent of its arable land the precipitation is less than 400 millimeters a year; on only one per cent of its arable land is the precipitation over 700 millimeters. In the USA more than 50 per cent of the arable land gets the most favorable precipitation for farming--700 millimeters per year. The best-endowed granary of the Soviet Union in this respect is Krasnodar Territory, or the Kuban (it takes its name from the river). This region gets a total of 450 millimeters of rain a year. In most grain-growing areas, and particularly in the Volga region (where the most valuable hardy wheat, including durum, is raised), the annual precipitation is even lighter--250-350 millimeters or less--and, moreover, it is poorly distributed in terms of growing months. As for the country's eastern granaries, Siberia and North Kazakhstan, their short growing season and poor weather conditions make them the least promising in the world.

More than two-thirds of the Soviet Union's grain is raised in areas which are hit regularly by droughts. The Ninth Five-Year Plan period (1971-1975) was unprecedented in this respect. Four out of five years were dry, and, according to meteorologists, the weather in 1975 was the worst since 1891. Nevertheless, the average annual yield in 1971-1975 topped 180 million tons, which is 13 million more than in 1966-1970 and 51 million more than in 1961-1965. Though we did not meet the targets of the Ninth Five-Year Plan, rising agricultural production under such conditions (which, meteorologists estimate, occur only once in a hundred years) can be considered a sizable achievement.

Far from being accidental, this growth resulted from the expanding supply of tractors and other timesaving machinery, the intensive reclamation program and use of chemicals and the outstanding work of Soviet plant breeders. A comprehensive approach to farm problems has made steady growth possible despite the whims of nature. That same approach predicates a higher grain yield, 215-220 million tons, in the Tenth Five-Year Plan period. Why not 250 million tons?

Even with Soviet weather conditions, such a yield is still possible. Then what prevents us from attaining it today? And what will be accomplished in the Tenth Five-Year Plan period to bring this optimal level closer?

Power Capacity per Farm Worker to Increase

It has been estimated that at least a 400-horsepower capacity is required per 100 hectares of plowland to achieve comprehensive mechanization in all areas of farming, so that operations can be handled in the shortest possible time. Soviet farmers have only 50 per cent of this capacity. Keep in mind, however, that farm-machinery building is one of the younger branches of Soviet industry. Fifty years ago our farms had only 27,000 low-powered tractors, 700 trucks and 2 (!) grain harvesters. The total capacity of all mechanical equipment was only 1.1 million horsepower as compared with today's more than 422 million horsepower. The growing capital investments in agriculture are responsible for the difference. Over the next five years a total of 171.7 billion rubles is to be allocated for farm development. This is equal to 80 percent of the investments in this area over the past 10 years. A considerable part of the sum will be spent on housing and social welfare and on additional machinery for collective farms. They will be provided with 1.9 million tractors, 1.35 million trucks, and 538,000 grain harvesters and other machines. An increase in the electric power available per farm worker will make it possible to intensify this branch. The growth of agricultural production in the Tenth Five-Year Plan period is to be achieved exclusively by higher labor productivity.

The income of collective farmers will rise on the average by 24-27 per cent as a result of increased production and higher labor productivity.

Reserves

The older generation of American and Canadian farmers remembers the mid-thirties, when the wheat harvest in the United States went down from 9.9 to 8.7 metric centners per hectare. In Canada it dropped by nearly 50 per cent as compared with the beginning of the century, from 13.3 to 7.1 metric centners per hectare. Soil exhaustion made us turn to agricultural chemistry, and the advantages of using fertilizer quickly became evident.

The intensive application of chemicals to agriculture began here in the sixties, at least 25 years later than in Canada and the United States. This lag was due to the inadequate industrial potential inherited from the past and, even more, to postwar devastation--the result of the Hitlerite occupation during World War II. In 1960 Soviet collective and state farms received only 11 million tons of fertilizer. Ten years later the amount increased fourfold, and in 1975 more than sixfold. But even today there is a shortage of fertilizer, and this restricts the growth of crop yields. In 1980 the fertilizer supply will be 115 million tons.

The crucial element, of course, is the farmers themselves, how effectively they utilize the land and machinery and the findings of selectionists.

A large-scale reclamation program is under way. In 1971-1975 about 4.5 million hectares of farmland were irrigated, and 4.4 million hectares were drained. In the Tenth Five-Year Plan period approximately that much land again will be added to the total reclaimed.

However, even the use of this "golden fund" is not enough. I believe that a sort of psychological barrier must be overcome. Soviet farmers and state and collective farm managers must recognize that mechanization and reclamation as well as selection and the use of chemicals signify a qualitatively new stage in the development of modern agricultural production.

What do I mean by this? Some 10 to 12 years ago, the average yield of grain on most farms did not exceed 9-10 metric centners per hectare. For centuries "a hundred-pood yield"--16 metric centners per hectare--was cited in Russian folklore as an ideal of peasant well-being. According to present standards, it is little more than a modest yield. Many farms take in 50-60 metric centners per hectare, while field trials of new grades on irrigated lands have raised the ceiling to 100-110 metric centners. But farmers, having reached a twofold or threefold increase, have not yet realized that this is only the beginning.

The "Guidelines for the Development of the National Economy of the USSR for 1976-1980," after nationwide discussion, were studied and adopted by the Twenty-fifth Congress of the Communist Party of the Soviet Union. After an appropriate decision by the USSR Supreme Soviet, such guidelines become part of the current five-year plan. Thus, the agricultural targets that I mentioned have been thoroughly considered and verified. We are talking not simply about desirable or possible directions for the development of Soviet agriculture, but about a concrete program for guaranteed growth.

AND THE PEASANTS: THEN AND NOW

Lenin repeatedly stressed the need for higher labour productivity as an essential part of the programme for socialist transformation of society. In an article, A Great Beginning, which likened the moral and economic import of the voluntary effort of the masses to the tender shoots of the communist attitude to labour, he wrote:

"Capitalism created a productivity of labour unknown under serfdom. Capitalism can be utterly vanquished, and will be utterly vanquished by socialism creating a new and much higher productivity of labour. This is a very difficult matter and must take a long time; but it has been started, and that is the main thing."

Characteristically, labour productivity in the Russian countryside before the Revolution was very low and was developing very slowly.

With the advent of capitalism in Russia in the 19th century the growth of productivity in agriculture was somewhat accelerated to roughly one per cent a year.

In the years of Soviet rule the gross output of farming and animal husbandry has shot up more than 150 per cent, while the number of people engaged in farm production has been reduced by half. This means that labour productivity has more than quintupled, with an attending increase in the Soviet farmers' income--both personal and social.

The net income of the Soviet farmers has increased fourfold since the pre-war year of 1940. It should be pointed out that the incomes of collective farmers from the farm are not taxed.

The peasants' life in tsarist Russia was a bitter one--poverty, backwardness and uncertainty in the future. The Great October Socialist Revolution opened for them the way to a free and happy future. Enthusiasm for their work and confidence in their way of life are common features of Soviet collective farmers, the masters of their land.

The conditions of work of the farmers have also undergone fundamental changes. The wooden plough and harrow, the sickle and flail were the only farm implements in pre-revolutionary Russia. Draught animals were a rarity, and there was only one horse for two or three households.

Today nearly two million tractors, tens of thousands of grain harvesters, lorries and other up-to-date farm machines ply the fields of the country.

The cultural level of the rural population has risen immeasurably. Before the revolution 75 per cent of the peasants could not read or write. Today illiteracy has been wiped out. Every collective farm has at least one secondary school (ten- or eight-year). The state and collective farms allocate large sums of money for education.

The face of the village has changed completely during the years of Soviet power. Ramshackle huts of clay and straw have given way to modern homes so that today's villagers could enjoy all the urban comforts in a rural environment.

Today life in the village is unthinkable without a House or Palace of Culture, nurseries and kindergartens, shops and service centres.

The farmer's way of life is becoming more and more like that of the city dweller.

The rural population now has 300 thousand shops, 55 thousand dining rooms, restaurants and cafes, more than 44 thousand community service establishments, over 111 thousand cultural centres, over 25 thousand kindergartens and nurseries--something the peasant never dreamed of in the old days.

Now the typical man of the village is an educated, cultured and well-trained person with versatile interests. Incidentally, the number of specialists with a higher or secondary education in the countryside grew 16-fold between 1940 and 1970, and now totals 821 thousand.

There have been big changes in the countryside. It was no easy matter to ensure a prosperous life for many millions of Russian peasants, mainly poor peasants, and give them access to culture.

All farmers irrespective of their nationality, have long since freed themselves of poverty. They are paid equally for equal work and their life keeps getting better.

As Soviet society grows and develops it is faced with new tasks. Since in terms of amenities village life is not yet up to the city standards, a good deal of attention is devoted to modernising and improving the comforts of the rural home. For example, during the 1966-70 plan period the number of daytime secondary schools in rural areas grew from 17.6 thousand to 25 thousand. More than three thousand public libraries and more than five thousand cultural centres were set up. Under the 1971-75 five-year plan about 10,000 new village schools, more than 9,000 public libraries and 13,000 cultural centres have been scheduled for construction. The amount of housing to be built will exceed that in the previous five-year plan period by 40 per cent.

The Soviet state extends 15-year credits to farmers for the improvement of community services and cultural amenities.

On an ever-growing scale construction in the countryside is carried out by centralised building organisations--interfarm establishments and big specialised state corporations. Altogether, over 2,000 interfarm corporations and more than 150 state corporations took care of construction in the countryside in 1973. They have about a thousand mobile mechanised units equipped with up-to-date machinery.

Modern city-type houses are going up in hundreds of villages near Moscow, in the Ukraine, Byelorussia, the Baltic republics, Central Asia and other parts of the Soviet Union, utilising the standard projects and specially developed designs.

TOWARD FURTHER IMPROVEMENTS: THE ROAD AHEAD

Over the past several years the emphasis has increasingly been on the intensive development of agriculture. This means that in the future agricultural production will grow not so much through the expansion of the area under crops as through a steady increase in crop yields.

The collective and state farms are seeking to make land cultivation more effective. While natural fertility is the result of long processes wrought by nature itself, induced fertility (as an addition to natural fertility) is the making of man equipped with machinery and know-how.

Before the socialist revolution the peasant knew no other fertilizer except the manure he gathered in his backyard. But even if he had a horse or cow, he could hardly expect to store up enough manure for his whole field, small though it was. Only a big landowner with a large dairy, or stud farm had the required quantity of manure. At present, farming is unthinkable without the use of chemicals which destroy pests and stimulate the growth of plants, especially vegetables. Plans are now used on a much wider scale than ever before for spreading fertilizers and pesticides on farm fields.

The USSR is the world's second largest producer of chemicals for agricultural needs. . . .

Soil amelioration work has been carried on in various parts of the Soviet Union on a wide scale. The area of irrigated and drained land in the USSR has increased by more than 150 per cent as compared with the pre-war period. . . .

The effect of the land improvement measures is obvious: though drained and irrigated land occupies only 6.5 per cent of the country's ploughland, it yields 24 per cent of its agricultural produce. . . .

Irrigation and development of desert land is proceeding apace, mainly in the republics of Central Asia and the Transcaucasia, in the drylands along the River Volga and in the southern Ukraine, in the North Caucasus and Moldavia.

One of the largest amelioration projects in the Ukraine is the North Crimean Canal which flows out of the Kakhovskoye reservoir on the Dnieper River. The canal will bring 165,000 hectares of drylands under irrigation. The other reservoirs of the Dnieper will irrigate over 400,000 hectares of land.

Work is under way on the amelioration of the marshlands which cover a large area of the Russian Federation, Byelorussia, the Baltic republics, and of the Ukraine.

In March 1974, the CPSU Central Committee and the Council of Ministers of the USSR have adopted a decision "On the Development of Agriculture in the Non-Chernozem Zone of the Russian Federation." This is in fact a programme for the all-round economic, social and cultural development of a huge territory--the entire central and northern part of the country. It will comprise ten per cent of the land used for agricultural purposes, 14 per cent of the land under plough, and it will be inhabited by 20 per cent of the population of the country. The implementation of this programme will not only change the face of this area, but will also have a significant effect on the country's economy. The state has already allocated 35,000 million roubles for the programme during the period 1976-80. This decision which sets long-term objectives up to 1990, is an example of the many-sided approach of the CPSU to the solution of large-scale national economic problems.

Socialism has radically changed the position of the farmer, the conditions of his life and work. The more mechanised the farming operations become and the more technically and mechanically-minded the farmer himself grows, as we showed earlier in this account, the smaller becomes the gap between industrial and agricultural work.

At the same time another, equally important historical process is taking place: life in the village is improving all the time, both materially and culturally, and this means that the standards of living in the village are coming closer to those of the city with all the comforts and benefits of civilisation.

Soviet agriculture . . . has come a long way, overcoming great difficulties and achieving spectacular results in the past 50 years.

ABOUT THE EDITOR AND THE CONTRIBUTORS

HARRY G. SHAFFER is Professor of Economics and Slavic and Soviet Area Studies at the University of Kansas. A New York University Ph.D., he has visited and carried out extensive interviews in the Soviet Union and all the socialist countries of East Europe except Albania.

Dr. Shaffer has published widely in the field of Slavic and Soviet Area Studies. He has to his credit six books and more than 40 articles and book reviews which have been published or reprinted in seven languages in such journals as The American Economic Review, The Journal of Industrial Economics, The Journal of Higher Education, The Slavic Review, Problems of Communism, The Russian Review, East Europe, Osteuropa, Osteuropa Wirtschaft, Revue de l'Est, The Antioch Review, Soviet and Eastern European Foreign Trade, The Kansas Business Review, and Queen's Quarterly. He has given papers and lectures at numerous professional conferences and at universities in the United States, Canada, and Germany.

M. GARDNER CLARK is Professor of Economics and Industrial and Labor Relations and a member of the Committee on Soviet Studies at Cornell University. He received his Ph.D. from Harvard University. He has been to the Soviet Union several times and visited collective farms in the USSR and Eastern Europe. Professor Clark has published three books and a number of articles on Soviet industry and agriculture.

VICTOR PERLO is an economic consultant, lecturer, and writer on economic affairs. He has published ten books, all of which have been translated into foreign languages. Several of them, and many of his numerous articles, deal with socialist countries. He has contributed articles to the American Economic Review, ACES Bulletin, Quarterly Review of Business and Economics, Political Affairs, Science & Society, New World Review, and to numerous economic and other scientific journals abroad. He has visited most of the socialist countries and talked there with the leading economic officials. He has taught courses on comparative economics in several colleges, most recently at the New School for Social Research in New York. Mr. Perlo is a Foreign Member of the Academy of Sciences of the German Democratic Republic.

RELATED TITLES
Published by
Praeger Special Studies

AGRICULTURE IN THE PEOPLE'S REPUBLIC OF CHINA: Structural Changes and Technical Transformation
Leslie T. C. Kuo

ECONOMIC DEVELOPMENT IN THE SOVIET UNION AND EASTERN EUROPE, Volume 1: Reforms, Technology, and Income Distribution
edited by
Zbigniew M. Fallenbuchl

ECONOMIC DEVELOPMENT IN THE SOVIET UNION AND EASTERN EUROPE, Volume 2: Sectoral Analysis
edited by
Zbigniew M. Fallenbuchl

ENVIRONMENTAL MISUSE IN THE SOVIET UNION
edited by
Fred Singleton

PERIODICALS ON THE SOCIALIST COUNTRIES AND ON MARXISM: A New Annotated Index of English-Language Publications
Harry G. Shaffer

SOCIAL SCIENTISTS AND POLICY MAKING IN THE USSR
edited by
Richard B. Remnek

WOMEN'S WORK IN SOVIET RUSSIA: Continuity in the Midst of Change
Michael Paul Sacks

WORLD FOOD PROSPECTS AND AGRICULTURAL POTENTIAL
Marylin Chou, David P. Harmon, Jr., Herman Kahn, and Sylvan H. Wittwer